Heavenly Bodies: Film Stars and Society

British Film Institute Cinema Series
Edited by Ed Buscombe

The British Film Institute Cinema Series opens up a new area of cinema publishing with books that will appeal to people who are already interested in the cinema but want to know more, written in an accessible style by authors who have some authority in their field. The authors write about areas of the cinema where there is substantial popular interest, but as yet little serious writing, or they bring together for a wider audience some of the important ideas which have been developed in film studies in recent years.

Published:
Richard Dyer: **Heavenly Bodies: Film Stars and Society**
Thomas Elsaesser: **New German Cinema: A History**
Jane Feuer: **The Hollywood Musical**
Douglas Gomery: **The Hollywood Studio System**
Colin MacCabe: **Goddard: Images, Sounds, Politics**
Steve Neale: **Cinema and Technology**

Forthcoming:
Lucy Fischer: **Shot/Countershot: Women in Film**
Jill Forbes: **Contemporary French Cinema**
Maria Kornatowska: **Contemporary Polish Cinema**

Richard Dyer

HEAVENLY BODIES

FILM STARS AND SOCIETY

First published 1986 by
THE MACMILLAN PRESS LTD
Houndmills, Basingstoke, Hampshire RG21 2XS
and London
Companies and representatives
throughout the world

ISBN 0–333–29540–4 (hardcover)
ISBN 0–333–29541–2 (paperback)

A catalogue record for this book is available from the British Library.

Printed in Hong Kong

Reprinted 1992

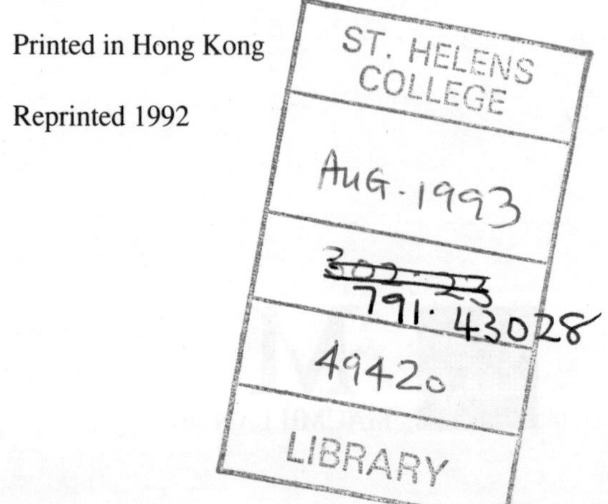

Contents

List of Illustrations

Cover photograph: Portrait of Joan Crawford by Eve Arnold, first published in **The Unretouched Woman** (New York: Knopf, 1976). © Eve Arnold

Preface

The studies in this book build on the work on stars that has emerged in the past few years. Writings and ideas about stars were surveyed in my previous book, **Stars** (Dyer, 1979a), and some of the ideas in that were extended and debated by several authors in **Star Signs** (BFI Education, 1982). The introduction to the latter by Christine Gledhill lays out very clearly the terms of disagreement in the whole area and what is at stake in them. The methods of analysis lying behind the work presented here are explained in my guide, **The Stars** (Dyer, 1979b).

The main purpose of this volume is to suggest by example how general ideas about the stars can be thought through in particular cases. Three issues emerging from previous work on them have especially informed this book. First, the need to find a way of understanding the social significance of stars which fully respects the way they function as media texts, yet does not fall into a view of a given star as simply reflecting some aspect of social reality that the analyst cared to name. It is through the concept of discourses that I have sought to bring together the star seen as a set of media signs with the various ways of understanding the world which influenced how people felt about that star.

Secondly, a major gap in work done on the stars, and indeed in media and cultural studies generally, is the role of the audience. I have sought to redress this by talking about how people could make sense of the star, the different ways in which he or she was read. The chapter on Garland is fully focused on this concern, since it is entirely concerned with how a particular subculture read a particular star, but the chapters on Monroe and Robeson also develop the idea that there were different audience readings of these two stars and that something of the range of these readings needs to be understood. I am working throughout with the idea of shared ways of reading, not with individual readings (or totted-up individual readings, the basic approach of audience surveys).

Thirdly, stars relate to very general ideas about society and the individual, ideas that I have sketched in the opening chapter. Much of the general impulse behind the interest in these ideas is political, stemming equally from two sources – on the one hand, a recognition of the failure of traditional left politics to think hard and feelingly enough about the human person, resulting in a tendency to lurch between a mushy humanism and the inhumanity of most communist regimes; and on the other hand, the growth of feminist and gay politics that have put directly on to the political agenda questions of emotions, sexuality, everyday life and more generally what being a person is and can be. Some of the ideas that stem from these political concerns are explored in Chapter 1.

A further political concern underpinning the book is the issue of representation. It has become increasingly clear that at all levels how we think and feel we are, how we are treated, is bound up with how we are represented as being. This becomes a directly political issue when groups decide that they do not accept or else wish to change the way they are represented. The choice of stars for extended study in this book relates to this concern. Monroe and Robeson are key figures in any debate about the representation of women and whiteness, blackness and masculinity. Garland, on the other hand, leads to a discussion of the culture produced by gay men, thus telling us not about how gay men have been represented by the dominant media but how gay men have used an image in the dominant media as a means of speaking to each other about themselves. All three are important because of the political value that has been claimed for them – Monroe has been seen as an exemplary figure of the situation of women in patriarchal capitalism; Robeson consciously tried to use his fame in an anti-racist and socialist way; Garland raises the question of the worth of subcultural practices with no explicitly political intent. The three may thus be seen not only as case studies in the analysis of stars, but as contributions to thinking about such political issues as – how do we understand how a given group of people are placed by the media? What are the possibilities of political intervention in the media? How can people as audiences use the media? How important are subcultures?

Richard Dyer

Acknowledgements

No book is the product of one person, even if she or he gets their name above the title; and the older you get the more people you have to thank. Rather than list all those whose thoughts, ideas and even casual remarks have made this book, and rather than risk leaving some individuals out, I should just like to thank the following – the library, the National Film Archive Stills Collection, Jim Adams and projection staffs of the British Film Institute; Ed Buscombe and Roma Gibson for help in preparing the text for publication; David Meeker, Bill Everson, George Eastman House and MOMA for help with Robeson material; Simon Crocker (John Kobal Collection); and Christine Gledhill and Jean McCrindle for their comments on earlier versions of the Monroe chapter. I have presented the work in the following chapters in various forms to many different groups of people and always their discussions have fed back into the chapters.

The Robeson chapter is dedicated to Jaime Credle; the Garland chapter to Michael Walker and the late Drew Griffiths.

Richard Dyer

Introduction

Eve Arnold's portrait of Joan Crawford gathers into one image three dimensions of stardom. Crawford is before two mirrors, a large one on the wall, the other a small one in her hand. In the former we see the Crawford image at its most finished; she is reduced to a set of defining features: the strong jaw, the gash of a mouth, the heavy arched eyebrows, the large eyes. From just such a few features, an impressionist, caricaturist or female impersonator can summon up 'Joan Crawford' for us. Meanwhile, in the small mirror we can see the texture of the powder over foundation, the gloss of the lipstick, the pencilling of the eyebrows – we can see something of the means by which the smaller image has been manufactured.

Neatly, we have two Crawford reflections. The placing of the smaller one, central and in sharpest focus, might suggest that this is the one to be taken as the 'real' Crawford. Eve Arnold is known as a photographer committed to showing women 'as they really are', not in men's fantasies of them. This photo appears in her collection **The Unretouched Woman** (1976), the title proclaiming Arnold's aim; it is accompanied by the information that Crawford wanted Arnold to do the series of photos of her to show what hard work being a star was. The style and context of the photo encourage us to treat the smaller image as the real one, as do our habits of thought. The processes of manufacturing an appearance are often thought to be more real than the appearance itself – appearance is mere illusion, is surface.

There is a third Crawford in the photograph, a back view slightly less sharply in focus than the mirror images. Both the large and the small facial images are framed, made into pictures. The fact that the different mirrors throw back different pictures suggests the complex relationship between a picture and that of which it is a picture, something reinforced by the fact that both mirrors reflect presentation: making-up and decorating the face. Both mirrors return a version of the front of the vague, shadowy figure before them. Is this third Crawford the real one, the real person who was the occasion of the images? This back view of Crawford establishes her as very much there, yet she is beyond our grasp except through the partial mirror images of her. Is perhaps the smaller mirror image the true reflection of what the actual person of Crawford was really like, or can we know only that there was a real person inside the images but never really know her? Which is Joan Crawford, really?

We can carry on looking at the Arnold photo like this, and our mind can constantly shift between the three aspects of Crawford; but it is the three of them taken together that make up the phenomenon Joan Crawford, and it is the insistent question of 'really' that draws us in, keeping us on the go from one aspect to another.

Logically, no one aspect is more real than another. How we appear is no less real than how we have manufactured that appearance, or than the 'we' that is doing the manufacturing. Appearances are a kind of reality, just as manufacture and individual persons are. However, manufacture and the person (a certain notion of the person, as I'll discuss) are generally thought to be more real than appearance in this culture. Stars are obviously a case of appearance – all we know of them is what we see and hear before us. Yet the whole media construction of stars encourages us to think in terms of 'really' – what is Crawford really like? which biography, which word-of-mouth story, which moment in which film discloses her as she really was? The star phenomenon gathers these aspects of contemporary human existence together, laced up with the question of 'really'.

The rest of this chapter looks at this complex phenomenon from two angles – first, the constitutive elements of stars, what they consist of, their production; secondly, the notions of personhood and social reality that they relate to. These are not separate aspects of stardom, but different ways of looking at the same overall phenomenon. How anything in society is made, how making is organised and understood, is inseparable from how we think people are, how they function, what their relation to making is. The complex way in which we produce and reproduce the world in technologically developed societies involves the ways in which we separate ourselves into public and private persons, producing and consuming persons and so on, and the ways in which we as people negotiate and cope with those divisions. Stars are about all of that, and are one of the most significant ways we have for making sense of it all. That is why they matter to us, and why they are worth thinking about.

Making Stars

The star phenomenon consists of everything that is publicly available about stars. A film star's image is not just his or her films, but the promotion of those films and of the star through pin-ups, public appearances, studio hand-outs and so on, as well as interviews, biographies and coverage in the press of the star's doings and 'private' life. Further, a star's image is also what people say or write about him or her, as critics or commentators, the way the image is used in other contexts such as advertisements, novels, pop songs, and finally the way the star can become part of the coinage of everyday speech. Jean-Paul Belmondo imitating Humphrey Bogart in **A bout de souffle** is part of Bogart's image, just as anyone saying, in a mid-European accent, 'I want to be alone' reproduces, extends and inflects Greta Garbo's image.

Star images are always extensive, multimedia, intertextual. Not all these manifestations are necessarily equal. A film star's films are likely to have a privileged place in her or his image, and I have certainly paid detailed attention to the films in the analyses that follow. However, even this is complicated. In the case of Robeson, his theatre, recording and concert work were undoubtedly more highly acclaimed than his film work – he was probably better known as a singer, yet more people would have seen him in films than in the theatre or concert hall. Later, in the period not covered here, he became equally important as a political activist. Garland became more important in her later years as a music hall, cabaret and recording star, although, as I argue in the Garland chapter, that later reputation then sent people back to her old films with a different kind of interest. Again, Monroe may now have become before everything else an emblematic figure, her symbolic meaning far outrunning what actually happens in her films.

As these examples suggest, not only do different elements predominate in different star images, but they do so at different periods in the star's career. Star images have histories, and histories that outlive the star's own lifetime. In the chapters that follow I have tried to reconstruct something of the meanings of Robeson and Monroe in the period in which they were themselves still making films – I've tried to situate them in relation to the immediate contexts of those periods. Robeson and Monroe have continued to be ethnic and sexual emblems as they were in their lifetime, but I have wanted to situate them in relation to the specific ways of understanding and feeling ethnic and sexual questions which were available in the thirties and fifties respectively, rather than in relation to what they mean in those terms now, although this would be an equally proper enquiry. (I did not, by the way, put ethnic and sexual in relation to Robeson and Monroe 'respectively', because Robeson is importantly situated in relation to ideas of sexuality just as Monroe is a profoundly ethnic image.) With Garland I have done the opposite – I have tried to look at her through a particular world-view, that of the white urban male gay subculture that developed in relation to her after her major period of film stardom and as she was becoming better known as a cabaret, recording and television star (and subject of scandal). The studies of Monroe, Robeson and Garland that follow are partial and limited, not only in the usual sense that all analyses are, but in being deliberaely confined to particular aspects of their images, at particular periods and with a particular interest in seeing how this is produced and registered in the films.

Images have to be made. Stars are produced by the media industries, film stars by Hollywood (or its equivalent in other countries) in the first instance, but then also by other agencies with which Hollywood is connected in varying ways and with varying degrees of influence. Hollywood controlled not only the stars' films but their promotion, the pin-ups and glamour portraits, press releases and to a large extent the fan clubs. In turn, Hollywood's connections with other media industries meant that what got into the press, who got to interview a star, what clips were released to television was to a large extent decided by Hollywood. But this is to present the process of star making as uniform and oneway. Hollywood, even within its own boundaries, was much more complex and contradictory than this. If there have always been certain key individuals in controlling positions (usually studio bosses and major producers, but also some directors, stars and other figures) and if they all share a general professional ideology, clustering especially around notions of entertainment, still Hollywood is also characterised by internecine warfare between departments, by those departments getting on with their own thing in their own ways and by a recognition that it is important to leave spaces for individuals and groups to develop their own ideas (if only because innovation is part of the way that capitalist industries renew themselves). If broadly everyone in Hollywood had a sense of what the Monroe, Robeson and Garland images were, still different departments and different people would understand and inflect the image differently. This already complex image-making system looks even more complex when one brings in the other media agencies involved, since there are elements of rivalry and competition between them and Hollywood, as well as co-operation and mutual influence. If the drift of the image emanates from Hollywood, and with some consistency within Hollywood, still the whole image-making process within and without Hollywood allows for variation, inflection, and contradiction.

What the audience makes of all this is something else again – and, as I've already suggested, the audience is also part of the making of the image. Audiences cannot make media images mean anything they want to, but they can select from the complexity of the image the meanings and feelings, the variations, inflections and contradictions, that work for them. Moreover, the agencies of fan magazines and clubs, as well as box office receipts and audience research, mean that the audience's ideas about a star can act back on the media producers of the star's image. This is not an equal to-and-fro – the audience is more disparate and fragmented, and does not itself produce centralised, massively available media images; but the audience is not wholly controlled by Hollywood and the media, either. In the case, for example, of feminist readings of Monroe (or of John Wayne) or gay male readings of Garland (or Montgomery Clift), what those particuar audiences are making of those stars is tantamount to sabotage of what the media industries thought they were doing.

Stars are made for profit. In terms of the market, stars are part of the way films are sold. The star's presence in a film is a promise of a certain kind of thing that you would see if you went to see the film. Equally, stars sell newspapers and magazines, and are used to sell toiletries, fashions, cars and almost anything else.

This market function of stars is only one aspect of their economic importance. They are also a property on the strength of whose name money can be raised for a film; they are an asset to the person (the star him/herself), studio and agent who controls them; they are a major part of the cost of a film. Above all, they are part of the labour that produces film as a commodity that can be sold for profit in the market place.

Stars are involved in making themselves into commodities; they are both labour and the thing that labour produces. They do not produce themselves alone. We can distinguish two logically separate stages. First, the person is a body, a psychology, a set of skills that have to be mined and worked up into a star image. This work, of fashioning the star out of the raw material of the person, varies in the degree to which it respects what artists sometimes refer to as the inherent qualities of the material; make-up, coiffure, clothing, dieting and body-building can all make more or less of the body features they start with, and personality is no less malleable, skills no less learnable. The people who do this labour include the star him/herself as well as make-up artistes, hairdressers, dress designers, dieticians, body-building coaches, acting, dancing and other teachers, publicists, pin-up photographers, gossip columnists, and so on. Part of this manufacture of the star image takes place in the films the star makes, with all the personnel involved in that, but one can think of the films as a second stage. The star image is then a given, like machinery, an example of what Karl Marx calls 'congealed labour', something that is used with further labour (scripting, acting, directing, managing, filming, editing) to produce another commodity, a film.

How much of a determining role the person has in the manufacture of her or his image and films varies enormously from case to case and this is part of the interest. Stars are examples of the way people live their relation to production in capitalist society. The three stars examined in subsequent chapters all in some measure revolted against the lack of control they felt they had – Robeson by giving up feature film-making altogether, Monroe by trying to fight for better parts and treatment, Garland by speaking of her experiences at MGM and by the way in which her later problems were credited to the Hollywood system. These battles are each central parts of the star's image and they enact some of the ways the individual is felt to be placed in relation to business and industry in contemporary society. At one level, they articulate a dominant experience of work itself under capitalism – not only the sense of being a cog in an industrial machine, but also the fact that one's labour and what it produces seem so divorced from each other – one labours to produce goods (and profits) in which one either does not share at all or only in the most meagre, back-handed fashion. Robeson's, Monroe's, Garland's sense that they had been used, turned into something they didn't control is particularly acute because the commodity they produced is fashioned in and out of their own bodies and psychologies.

Other stars deliver different stories, of course. June Allyson, in interviews and in her biography 'with Frances Spatz-Leighton', sings the praises of the job security provided by the studio system, of big capital, just as in her movies she perfected the role of the happy stay-at-home housewife who saw it as her role to support her man in his productive life, whether he produced music (as in **The Glenn Miller Story**) or profits (as in **Executive Suite**). There is a consistency between her 'contented housewife' screen image, her satisfaction with her working conditions, the easygoing niceness in the tone of the biography and interviews. She thus represents the possibility of integrated, mutually supporting spheres of life, not the tension between screen image, manufacture and real person that Monroe, Garland and Robeson suggest.

Many male stars – Clark Gable, Humphrey Bogart, Paul Newman, Steve McQueen – suggest something else again. In each, sporting activity is a major – perhaps the major – element in their image; they are defined above all as people for whom having uncomplicated fun is paramount, and this is implicitly carried over into their reported attitude to their work. But equally work isn't important, it's just something you do so as to have the wherewithal to play polo, sail yachts, race cars. This is, then, an instrumental attitude towards manufacture, not the antagonistic one of Garland, Robeson and Monroe, nor the integrated one of Allyson, nor yet again the committed one of, for example, Fred Astaire, Joan Crawford or Barbra Streisand. These last three suggest different relations of commitment to work – Astaire to technical mastery, in the endless stories of his perfectionist attitude towards rehearsal and the evidence of it on the screen; Crawford in her total slogging away at all aspects of her image and her embodiment of the ethic of hard work in so many of her films; Streisand in her control over the films and records she makes, a reported shopfloor control that also shows in the extremely controlled and detailed nature of her performance style. Whatever the particular inflection, stars play out some of the ways that work is lived in capitalist society. My selection of Monroe, Robeson and Garland is different only in that there is in them an element of protest about labour under capitalism which you do not find in Allyson, Gable, Astaire, Streisand and the rest.

The protests of Robeson, Monroe and Garland are individual protests. Robeson and Monroe could be taken as protests emblematic of the situation of black people and women respectively, and have been properly used as such. But they remain individualised, partly because the star system is about the promotion of the individual. Protest about the lack of control over the outcome of one's labour can remain within the logic of individualism. The protests of Robeson, Monroe and Garland are of the individual versus the anomic corporation; they are protests against capitalism that do not recognise themselves as such, protests with deep resonances within the ideologies of entrepreneurial capitalism. They speak in the name of the individual and of the notion of success, not in the name of the individual as part of a collective organisation of labour and production. (Robeson alone began to move in that direction in his ensemble theatre work, and in his deliberately emblematic role in political activity in later years.)

A star image consists both of what we normally refer to as his or her 'image', made up of screen roles and obviously stage-managed public appearances, and also of images of the manufacture of that 'image' and of the real person who is the site or occasion of it. Each element is complex and contradictory, and the star is all of it taken together. Much of what makes them interesting is how they articulate aspects of living in contemporary society, one of which, the nature of work in capitalist society, I've already touched on. In the chapters that follow I want to look at the ways in which three particular stars relate to three aspects of social life – sexuality, ethnicity and sexual identity. Even being that specific, it is still complicated, I'm still wanting to keep some sense of the multiplicity of readings even of those stars in those terms. In the rest of this chapter, however, I want to risk even wider generalisations. Work, sexuality, ethnicity and sexual identity themselves depend on more general ideas in society about what a person is, and stars are major definers of these ideas.

Living Stars

Stars articulate what it is to be a human being in contemporary society; that is, they express the particular notion we hold of the person, of the 'individual'. They do so complexly, variously – they are not straightforward affirmations of individualism. On the contrary, they articulate both the promise and the difficulty that the notion of individuality presents for all of us who live by it.

'The individual' is a way of thinking and feeling about the discrete human person, including oneself, as a separate and coherent entity. The individual is thought of as separate in the sense that she or he has an existence apart from anything else – the individual is not just the sum of his or her social roles or actions. He or she may only be perceived through these things, may even be thought to be formed by them, yet there is, in this concept of the person, an irreducible core of being, the entity that is perceived within the roles and actions, the entity upon which social forces act. This irreducible core is coherent in that it is supposed to consist of certain peculiar, unique qualities that remain constant and give sense to the person's actions and reactions. However much the person's circumstances and behaviour may change, 'inside' they are still the same individual; even if 'inside' she or he has changed, it is through an evolution that has not altered the fundamental reality of that irreducible core that makes her or him a unique individual.

At its most optimistic, the social world is seen in this conception to emanate from the individual, and each person is seen to 'make' his or her own life. However, this is not necessary to the concept. What is central is the idea of the separable, coherent quality, located 'inside' in consciousness and variously termed 'the self', 'the soul', 'the subject' and so on. This is counterposed to 'society', something seen as logically distinct from the individuals who compose it, and very often as inimical to them. If in ideas of 'triumphant individualism' individuals are seen to determine society, in ideas of 'alienation' individuals are seen as cut adrift from and dominated, battered by the anonymity of society. Both views retain the notion of the individual as separate, irreducible, unique.

It is probably true to say that there has never been a period in which this concept of the individual was held unproblematically throughout society. The notion of the individual has always been accompanied by the gravest doubts as to its tenability. It is common, for instance, to characterise Enlightenment philosophy as one of the most shiningly optimistic assertions of individuality; yet two of its most sparkling works, Hume's **An Essay on Human Understanding** and Diderot's **Rameau's Nephew,** fundamentally undercut any straightforward belief in the existence of the coherent, stable, inner individual; Hume by arguing that all we can know as our self is a series of sensations and experiences with no necessary unity or connection, Diderot by focusing on the vital, theatrical, disjointed character of Rameau's nephew, so much more 'real' than Diderot, the narrator's stodgily maintained coherent self.

If the major trend of thought since the Renaissance, from philosophical rumination to common sense, has affirmed the concept of the individual, there has been an almost equally strong counter-tradition of ideas that have severely dented our confidence in ourselves: Marxism, with its insistence that social being determines consciousness and not vice versa, and, in its economist variant, with its vision of economic forces propelling human events forward; psychoanalysis, with its radical splitting of consciousness into fragmentary, contradictory parts; behaviourism, with its view of human beings controlled by instinctual appetites beyond consciousness; linguistics and models of communication in which it is not we who speak language, but language which speaks us. Major social and political developments have been understood in terms of the threat they pose to the individual: industrialisation can be seen to have set the pace for a whole society in which people are reduced to being cogs in a machine;

totalitarianism would seem to be the triumph, easily achieved, of society over the individual; the development of mass communications, and especially the concomitant notion of mass society, sees the individual swallowed up in the sameness produced by centralised, manipulative media which reduce everything to the lowest common denominator. A major trajectory of twentieth-century high literature has examined the disintegration of the person as stable ego, from the fluid, shifting self of Woolf and Proust to the minimal self of Beckett and Sarraute. 'Common sense' is no less full of tags acknowledging this bruised sense of self: the sense of forces shaping our lives beyond our control, of our doing things for reasons that we don't understand, of our not recognising ourself in actions we took yesterday (to say nothing of years ago), of not seeing ourselves in photographs of ourselves, of feeling strange when we recognise the routinised nature of our lives – none of this is uncommon.

Yet the idea of the individual continues to be a major moving force in our culture. Capitalism justifies itself on the basis of the freedom (separateness) of anyone to make money, sell their labour how they will, to be able to express opinions and get them heard (regardless of wealth or social position). The openness of society is assumed by the way that we are addressed as individuals – as consumers (each freely choosing to buy, or watch, what we want), as legal subjects (free and responsible before the law), as political subjects (able to make up our mind who is to run society). Thus even while the notion of the individual is assailed on all sides, it is a necessary fiction for the reproduction of the kind of society we live in.

Stars articulate these ideas of personhood, in large measure shoring up the notion of the individual but also at times registering the doubts and anxieties attendant on it. In part, the fact that the star is not just a screen image but a flesh and blood person is liable to work to express the notion of the individual. A series of shots of a star whose image has changed – say, Elizabeth Taylor – at various points in her career could work to fragment her, to present her as nothing but a series of disconnected looks; but in practice it works to confirm that beneath all these different looks there is an irreducible core that gives all those looks a unity, namely Elizabeth Taylor. Despite the elaboration of roles, social types, attitudes and values suggested by any one of these looks, one flesh and blood person is embodying them all. We know that Elizabeth Taylor exists apart from all these looks, and this knowledge alone is sufficient to suggest that there is a coherence behind them all.

It can be enough just to know that there was one such person, but generally our sense of that one person is more vivid and important than all the roles and looks s/he assumes. People often say that they do not rate such and such a star because he or she is always the same. In this view, the trouble with, say, Gary Cooper or Doris Day, is that they are always Gary Cooper and Doris Day. But if you like Cooper or Day, then precisely what you value about them is that they are always 'themselves' – no matter how different their roles, they bear witness to the continuousness of their own selves.

This coherent continuousness within becomes what the star 'really is'. Much of the construction of the star encourages us to think this. Key moments in films are close-ups, separated out from the action and interaction of a scene, and not seen by other characters but only by us, thus disclosing for us the star's face, the intimate, transparent window to the soul. Star biographies are devoted to the notion of showing us the star as he or she really is. Blurbs, introductions, every page assures us that we are being taken 'behind the scenes', 'beneath the surface', 'beyond the image', there where the truth resides. Or again, there is a rhetoric of sincerity or authenticity, two qualities greatly prized in stars because they guarantee, respectively, that the star really means what he or she says, and that the star really is what she or he appears to be. Whether caught in the unmediated moment of the close-up, uncovered by the biographer's display of ruthless uncovering, or present in the star's indubitable sincerity and authenticity, we have a privileged reality to hang on to, the reality of the star's private self.

The private self is further represented through a set of oppositions that stem from the division of the world into private and public spaces, a way of organising space that in turn relates to the idea of the separability of the individual and society:

private	public
individual	society
sincere	insincere
country	city
small town	large town
folk	urban
community	mass
physical	mental
body	brain

naturalness	artifice
sexual intercourse	social intercourse
racial	ethnic

When stars function in terms of their assertion of the irreducible core of inner individual reality, it is generally through their associations with the values of the left-hand column. Stars like Clark Gable, Gary Cooper, John Wayne, Paul Newman, Robert Redford, Steve McQueen, James Caan establish their male action-hero image either through appearing in Westerns, a genre importantly concerned with nature and the small town as centres of authentic human behaviour, and/or through vivid action sequences, in war films, jungle adventures, chase films, that pit the man directly, physically against material forces. It is interesting that with more recent examples of this type – Clint Eastwood, Harrison Ford – there has been a tendency either to give their films a send-up or tongue-in-cheek flavour (Eastwood's chimp films, Ford as Indiana Jones) or else a hard, desolate, alienated quality (Eastwood in **Joe Kidd,** Ford in **Blade Runner**), as if the values of masculine physicality are harder to maintain straight-facedly and unproblematically in an age of microchips and a large scale growth (in the USA) of women in traditionally male occupations.

The private self is not always represented as good, safe or positive. There is an alternative tradition of representing the inner reality of men, especially, which stretches back at least as far as the romantic movement. Here the dark, turbulent forces of nature are used as metaphors for the man's inner self: Valentino in **The Son of the Sheik,** the young Laurence Olivier as Heathcliff in **Wuthering Heights** and as Maxim de Winter in **Rebecca.** In the forties and fifties the popularisation of psychoanalysis added new terms to the private:public opposition. Thus:

private	public
subconscious	conscious
Id	Ego

and in the still more recent Lacan inflection:

Imaginary	Symbolic

These have been particularly important in the subsequent development of male stars, where the romantic styles of brooding, introspective, mean-but-vulnerable masculinity have been given Oedipal, psychosexual, paranoid or other crypto-psycho-analytical inflections with stars like Montgomery Clift, James Dean, Marlon Brando, Anthony Perkins, Jack Nicholson, Richard Gere. Recent black male stars such as Jim Brown, Richard Roundtree and Billy Dee Williams are interesting in that their fiercely attractive intensity seems closer to the 'dangerous' romantic tradition proper; at the same time they also draw on the old stereotype of the black man as brute, only now portraying this as attractive rather than terrifying; and they are almost entirely untouched by the psychoanalytical project of rationalising and systematising and naming the life of the emotions and sensations. All these male stars work variations on the male inner self as negative, dangerous, neurotic, violent, but always upholding that as the reality of the man, what he is really like.

The stars analysed in the rest of this book also have strong links with the left-hand, 'private' column. Monroe was understood above all through her sexuality – it was her embodiment of current ideas of sexuality that made her seem real, alive, vital. Robeson was understood primarily through his racial identity, through attempts to see and, especially, hear him as the very essence of the Negro folk. Both were represented insistently through their bodies – Monroe's body was sexuality, Robeson's was the nobility of the black race. Garland too belongs with the left-hand column, initially through her roles as country or small-town girl, later through the way her body registered both her problems and her defiance of them. All the descriptions of her from her later period begin by describing the state of her body and speculating from that on what drugs, drink, work and temperament have done to it, and yet how it continues to be animated and vital. Not only are Monroe, Robeson, Garland stars who are thought to be genuine, who reveal their inner selves, but the final touchstone of that genuineness is the human body itself. Stars not only bespeak our society's investment in the private as the real, but also often tell us how the private is understood to be the recovery of the natural 'given' of human life, our bodies. Yet as the chapters that follow argue, what we actually come up against at this point is far from straightforwadly natural; it is particular, and even rather peculiar, ways of making sense of the body. The very notions of sexuality and race, so apparently rooted in the body, are historically and culturally specific ideas about the body, and it is these that Monroe and Robeson, especially, enact, thereby further endowing them with authenticity.

What is at stake in most of the examples discussed so far

is the degree to which, and manner in which, what the star really is can be located in some inner, private, essential core. This is how the star phenomenon reproduces the overriding ideology of the person in contemporary society. But the star phenomenon cannot help being also about the person in public. Stars, after all, are always inescapably people in public. If the magic, with many stars, is that they seem to be their private selves in public, still they can also be about the business of being in public, the way in which the public self is endlessly produced and remade in presentation. Those stars that seem to emphasise this are often considered 'mannered', and the term is right, for they bring to the fore manners, the stuff of public life. When such stars are affirmative of manners and public life they are often, significantly enough, European or with strong European connections – stars to whom terms like suave, gracious, debonair, sophisticated, charming accrue, such as Fred Astaire, Margaret Sullavan, Cary Grant, David Niven, Deborah Kerr, Grace Kelly, Audrey Hepburn, Rex Harrison, Roger Moore. These are people who have mastered the public world, in the sense not so much of being authentically themselves in it nor even of being sincere, as of performing in the world precisely, with poise and correctness. They get the manners right. An additional example might be Sidney Poitier, only with him the consummate ease of his public manners comes up against the backlog of images of black men as raging authenticities, with the result that in his films of the fifties and sixties he is not really able to be active in public, he is a good performer who doesn't perform anything. It is only with **In the Heat of the Night** that something else emerges, a sense of the tension attendant on being good in public, a quality that brings Poitier here into line with a number of other stars who suggest something of the difficulty and anxiety attendant upon public performance.

Many of the women stars of screwball comedy – Katharine Hepburn, Carole Lombard, Rosalind Russell, and more recently Barbra Streisand – have the uncomfortable, sharp quality of people who do survive and succeed in the public world, do keep up appearances, but edgily, always seen to be in the difficult process of doing so. Bette Davis's career has played variations on this representation of public performance. Many of her films of the thirties and forties exploit her mannered style to suggest how much her success or survival depends upon an ability to manipulate manners, her own and those of people around her, to get her own way (**Jezebel, The Little Foxes**), to cover her tracks out of courage (**Dark Victory**) or guilt (**The Letter**), to maintain a public presence at all costs for a greater good than her own (**The Private**

Life of Elizabeth and Essex), to achieve femininity (**Now Voyager**) and so on. If being in public for Davis in these films is hypertense, registered in her rapid pupil movements, clenching and unclenching fists, still in the thirties and forties she is enacting the excitement, the buzz of public life, of being a person in public. Later films become something like the tragedy of it. **All About Eve** details the cost of keeping up appearances, maintaining an image. **Whatever Happened to Baby Jane?** evokes the impossibility of achieving again the public role that made her character feel good. Yet the end of **Baby Jane** affirms the public self as a greater reality than the private self cooped up in the dark Gothic mansion – we learn that it is Crawford not Davis who is the baddie; away from the house, on the beach, surrounded by people, the ageing Jane can become the public self she really is, Baby Jane. Davis's career thus runs the gamut of the possibilities of the private individual up against public society; from, in the earlier films, triumphant individualism, the person who makes their social world, albeit agitatedly, albeit at times malignantly, to, in the later films, something like alienation, the person who is all but defeated by the demands of public life, who only hangs on by the skin of their teeth – until the up-tempo happy ending.

The private/public, individual/society dichotomy can be embodied by stars in various ways; the emphasis can fall at either end of the spectrum, although it more usually falls at the private, authentic, sincere end. Mostly too there is a sense of 'really' in play – people/stars are really themselves in private or perhaps in public but at any rate somewhere. However, it is one of the ironies of the whole star phenomenon that all these assertions of the reality of the inner self or of public life take place in one of the aspects of modern life that is most associated with the invasion and destruction of the inner self and corruptibility of public life, namely the mass media. Stars might even seem to be the ultimate example of media hype, foisted on us by the media's constant need to manipulate our attention. We all know how the studios build up star images, how stars happen to turn up on chat shows just when their latest picture is released, how many of the stories printed about stars are but titillating fictions; we all know we are being sold stars. And yet those privileged moments, those biographies, those qualities of sincerity and authenticity, those images of the private and the natural can work for us. We may go either way. As an example, consider the reactions at the time to John Travolta in **Saturday Night Fever.** I haven't done an audience survey, but people seemed to be fairly evenly divided. For those not taken with him, the incredible build-up to the film, the way you knew what his image was before you saw the film, the coy but blatant emphasis on his sex appeal in the film, the gaudy artifice of the disco scene, all merely confirmed him as one great phoney put-on on the mass public. But for those for whom he and the film did work, there were the close-ups revealing the troubled pain behind the macho image, the intriguing off-screen stories about his relationship with an older woman, the spontaneity (= sincerity) of his smile, the setting of the film in a naturalistically portrayed ethnic subculture. A star's image can work either way, and in part we make it work according to how much it speaks to us in terms we can understand about things that are important to us.

Nonetheless, the fact that we know that hype and the hard sell do characterise the media, that they are supreme instances of manipulation, insincerity, inauthenticity, mass public life, means that the whole star phenomenon is profoundly unstable. Stars cannot be *made* to work as affirmations of private or public life. In some cases, the sheer multiplicity of the images, the amount of hype, the different stories told become overwhelmingly contradictory. Is it possible still to have any sense of Valentino or Monroe, their persons, apart from all the things they have been made to mean? Perhaps, but at best isn't it a sense of the extraordinary fragility of their inner selves, endlessly fragmented into what everyone else, including us, wanted them to be? Or it may be that what interests us is the public face, accepting the artifice and fantasy for what it is – do we ask for sincerity and authenticity from Jayne Mansfield or Diana Ross, Groucho, Harpo or Chico Marx?

Or we may read stars in a camp way, enjoying them not for any supposed inner essence revealed but for the way they jump through the hoops of social convention. The undulating contours of Mae West, the lumbering gait and drawling voice of John Wayne, the thin, spiky smile of Joan Fontaine – each can be taken as an emblem of social mores: the ploys of female seduction, the certainty of male American power, the brittle niceness of upper class manners. Seeing them that way is seeing them as appearance, as image, in no way asking for them to be what they are, really.

On rare occasions a star image may promote a sense of the social constructedness of the apparently natural. The image of Lena Horne in her MGM films does this in relation to ideas of black and female sexuality. Her whole act in these films – and often it is no more than a turn inserted into the narrative flow of the film – promotes the idea of natural, vital sexuality, with her flashing eyes, sinuous arm movements and suggestive vocal delivery. That people saw this as the ultimate in unfettered feminine libido is widely attested, yet as an act it has an extraordinary quality, a kind of metallic sheen and intricate precision that suggests the opposite of animal vitality. In an interview with Michiko Kakutani in the **New York Times** (Sunday, 3 May 1981, section D, pp1,24), Lena Horne discussed her image in this period in relation to her strategy of survival in the period as a black women:

> *Afraid of being hurt, afraid of letting her anger show, she says she cultivated an image that distanced her from her employers, her colleagues, and from her audiences as well. If audience members were going to regard her as no more than an exotic performer – 'Baby, you sure can sing, but don't move next door' – well, then, that's all they'd get. By focusing intently on the notes and lyrics of a song, she was able to shut out the people who were staring at her, and over the years, she refined a pose of sophisticated aloofness, a pose that said, 'You're getting the singer, but not the woman.' 'I used to think, "I'm black and I'm going to isolate myself because you don't understand me,"' she says. 'All the things people said – sure, they hurt, and it made me retreat even further. The only thing between me and them was jive protection.'*

It is rare for a performer to understand and state so clearly both how they worked and the effect of it, but this catches exactly Horne's image in the forties and fifties, its peerless surface, its presentation of itself *as* surface, its refusal to corroborate, by any hint of the person giving her self, the image of black sexuality that was being wished on her. This could not, did not stop audiences reading her as transparently authentic sexuality; but it was some sort of strategy of survival that could also be seen for what it was, a denaturalising of the ideas of black sexuality.

I have been trying to describe in this chapter some of the ways in which being interested in stars is being interested in how we are human now. We're fascinated by stars because they enact ways of making sense of the experience of being a person in a particular kind of social production (capitalism), with its particular organisation of life into public and private spheres. We love them because they represent how we think that experience is or how it would be lovely to feel that it is. Stars represent typical ways of behaving, feeling and thinking in contemporary society, ways that have been socially, culturally, historically constructed. Much of the ideological investment of the star phenomenon is in the stars seen as individuals, their qualities seen as natural. I do not wish to deny that there are individuals, nor that they are grounded in the given facts of the human body. But I do wish to say that what makes them interesting is the way in which they articulate the business of being an individual, something that is, paradoxically, typical, common, since we all in Western society have to cope with that particular idea of what we are. Stars are also embodiments of the social categories in which people are placed and through which they have to make sense of their lives, and indeed through which we make our lives – categories of class, gender, ethnicity, religion, sexual orientation, and so on. And all of these typical, common ideas, that have the feeling of being the air that you breathe, just the way things are, have their own histories, their own peculiarities of social construction.

Because they go against the grain of the individualising, naturalising emphasis of the phenomenon itself, these insistences on the typical and social may seem to be entirely imported from theoretical reflection. Yet ideas never come entirely from outside the things they are ideas about, and this seems particularly so of the star phenomenon. It constantly jogs these questions of the individual and society, the natural and artificial, precisely because it is promoting ideas of the individual and the natural in media that are mass, technologically elaborated, aesthetically sophisticated. That central paradox means that the whole phenomenon is unstable, never at a point of rest or equilibrium, constantly lurching from one formulation of what being human is to another. This book is an attempt to tease out some of those formulations in particular cases, to see how they work, to get at something of the contradictions of what stars are, really.

Monroe and Sexuality

The denial of the body is delusion. No woman transcends her body.
Joseph C. Rheingold

Men want women pink, helpless and do a lot of deep breathing.
Jayne Mansfield

Stars matter because they act out aspects of life that matter to us; and performers get to be stars when what they act out matters to enough people. Though there is a sense in which stars must touch on things that are deep and constant features of human existence, such features never exist outside a culturally and historically specific context. So, for example, sexual intercourse takes place in all human societies, but what intercourse means and how much it matters alters from culture to culture, and within the history of any culture. The argument in this chapter is that, in the fifties, there were specific ideas of what sexuality meant and it was held to matter a very great deal; and because Marilyn Monroe acted out those specific ideas, and because they were *felt* to matter so much, she was charismatic, a centre of attraction who seemed to embody what was taken to be a central feature of human existence at that time.

My method is to read Monroe through the ideas about sexuality that circulated in the fifties, ideas that I centre on two strands, the one most forcefully represented by **Playboy** magazine, the other concerned with comprehending female sexuality. I want to use the term discourse for these strands, to indicate that we are not dealing with philosophically coherent thought systems but rather with clusters of ideas, notions, feelings, images, attitudes and assumptions that, taken together, make up distinctive ways of thinking and feeling about things, of making a particular sense of the world. A discourse runs across different media and practices, across different cultural levels – from the self-conscious Playboy 'philosophy' to the habitual forms of the pin-up, from psychoanalytic theory through psychotherapeutic practices to the imagery of popular magazines and best-selling novels.

To a large extent, the analysis which follows stays 'within' discourse. Though I do make some reference to the material world which discourse itself refers to, I don't really analyse it in any detail. As far as my argument goes, Monroe is charismatic because she embodies what the discourses designate as the important-at-the-time central features of human existence. In this way I want to avoid a simplistic correlation between Monroe and either the actual social structures of the fifties or the lived experience of 'ordinary' women and men. However, this is a limitation of my approach and I wouldn't want to give the impression that there is no correlation between discourses, structures and experiences. If the discourses (and Monroe) did not in fact have some purchase on how people lived in the social and economic conditions of their time, I do not see how they can have in any sense worked or been effective; I don't think people would have paid to go and see her. But I do not show here just what the precise nature of the connections between discourses, social structures and experiences are in this case.

In stressing the importance of sexuality in Marilyn Monroe's image, it might seem that I am just another commentator doing to Monroe what was done to her throughout her life, treating her solely in terms of sex. Perhaps that is a danger, but I hope that I am not just reproducing this attitude to Monroe but trying to understand it and historicise it. Monroe may have been a wit, a subtle and profound actress, an intelligent and serious woman; I've no desire to dispute this and it is important to recognise and recover those qualities against the grain of her image. But my purpose is to understand the grain itself, and there can be no question that this is overwhelmingly and relentlessly constructed in terms of sexuality. Monroe = sexuality is a message that ran all the way from what the media made of her in the pin-ups and movies to how her image became a reference point for sexuality in the coinage of everyday speech.

She started her career as a pin-up, and one can find no type of image more single-mindedly sexual than that. Pin-ups remained a constant and vital aspect of her image right up to her death, and the pin-up style also indelibly marked other aspects, such as public appearances and promotion for films. The roles she was given, how she was filmed and the reviews she got do little to counteract this emphasis.

She plays, from the beginning, 'the girl', defined solely by age, gender and sexual appeal. In two films, she does not even have a name (**Scudda Hoo! Scudda Hay!** 1948 and **Love Happy,** 1950) and in three other cases, her character has no biography beyond being 'the blonde' (**Dangerous Years,** 1948; **The Fireball,** 1950 and **Right Cross,** 1950). Even when any information about the character is supplied, it serves to reinforce the basic anonymity of the role. For instance, when the character has a job, it is a job that – while it may, like that of secretary, be in fact productive – is traditionally (or cinematically) thought of as being one where the woman is on show, there for the pleasure of men. These jobs in Monroe's early films are chorus girl (**Ladies of the Chorus,** 1948 and **Ticket to Tomahawk,** 1950), actress (**All About Eve,** 1950 – the film emphasises that the character has no talent) or secretary (**Home Town Story,** 1951; **As Young As You Feel,** 1951 and **Monkey Business,** 1952). There is very little advance on these roles in the later career. She has no name in **The Seven Year Itch** (1955), even in the credits she is just 'the Girl'. She is a chorus girl in **Gentlemen Prefer Blondes** (1953), **There's No Business Like Show Business** (1954), **The Prince and the Showgirl** (1957) and **Let's Make Love** (1960), and a solo artiste of no great talent in **River of No Return** (1954), **Bus Stop** (1956) and **Some Like It Hot** (1959). She is a model (hardly an extension of the role repertoire) in **How to Marry a Millionaire** (1953) and **The Seven Year Itch** (1955), and a prostitute in **O. Henry's Full House** (1952). Thus even in her prestige roles, **Bus Stop** and **The Prince and the Showgirl,** the social status of the person she plays remains the same (this does not mean, of course, that there is no difference between these characters or their portrayal). The tendency to treat her as nothing more than her gender reaches its peak with **The Misfits** (1961), where, from being the 'girl' in the early films, she now becomes the 'woman', or perhaps just 'Woman' – Roslyn has no biography, she is just 'a divorcee'; the symbolic structure of the film relates her to Nature, the antithesis of culture, career, society, history . . .

There is no question that Monroe did a lot with these roles, but it is nearly always against the grain of how they are written, and how they are filmed too. She is knitted into the fabric of the film through point-of-view shots located in male characters – even in the later films, and virtually always in the earlier ones, she is set up as an object of male sexual gaze. Frequently too she is placed within the frame of the camera in such a way as to stand out in silhouette, a side-on tits and arse positioning obsessively repeated throughout her films. One of the most sustained

Marilyn Monroe and Laurence Olivier in **The Prince and the Showgirl** (1957)

treatments of her as sexual spectacle is **The Prince and the Showgirl,** a film produced by her own company and directed by Sir Laurence Olivier, a film you might expect to be different in approach. Superficially it is – the lines more theatrically witty, the sets more tastefully dressed than in her 20th Century-Fox extravaganzas. Yet the film constantly plays with our supposed desire to see Monroe as sexual spectacle. The first few minutes of the film concerning the Monroe character, Elsie, are set backstage at 'The Cocoanut Girl', in and around the showgirls' dressing room. Such settings always raise voyeurs' hopes, and the film teases them. One shot follows the call boy along the passage to the dressing room that Elsie shares with the other girls; he knocks and enters, leaving the door open, but from where the camera is positioned we can't see the girls; after a moment, however, the camera cranes round so that it/we can see in on the girls, but the tease is also a cheat – they are all fully dressed. Later in the film Monroe and Olivier are posed on either side of the screen. Olivier is face-on, in a shapeless dressing-gown against a dark background of bookshelves – his figure is not clearly visible and the *mise-en-scène* identifies him with the intellect (books). Monroe is posed side-on, in a tight dress that facilitates another tits and arse shot as in earlier films. Behind her is a nude female statuette. Her figure is thrust at us, and the *mise-en-scène* identifies her with woman-as-body, woman-as-spectacle.

Hardly surprisingly, the reviewers also saw her over-

*All quotations of reviews taken from Conway and Ricci (1964), unless otherwise stated.

whelmingly in terms of sex.* Typical of the early period are descriptions of her as 'a beautiful blonde' in **The Asphalt Jungle** (1950), 'curvey Marilyn Monroe' in **As Young As You Feel,** as having a 'shapely chassis' and being 'a beautiful blonde' in **Let's Make It Legal** (1951). Just on the brink of full stardom (and after having, as we now suppose, made a mark, beyond that of sex object, in **The Asphalt Jungle** and **All About Eve**), a critic writes, à propos of **We're Not Married** (1952), 'Marilyn Monroe supplies the beauty at which she is Hollywood's currently foremost expert.' Barbara Stanwyck recalls the gentlemen of the press, when they visited the lot of **Clash By Night** (1952), announcing that they were not interested in her, Stanwyck, the star of the film – 'We don't want to speak to her. We know everything about her. We want to talk to the girl with the big tits.' Again, even as late as **Some Like It Hot** and **Let's Make Love,** the same kind of remarks are found among the reviews – of the former: ' . . . Miss Monroe, whose figure simply cannot be overlooked . . .' and of the latter: ' . . . the famous charms are in evidence'. Thus the direct physical presence of Monroe is never lost sight of behind other later emphases, such as her wit or acting abilities, though it is true that there is a certain jokey defensiveness about much of the later reviews' harping on sex appeal, as if in acknowledgement of the other claims made for Monroe in the period.

Given this emphasis in the pin-ups, movies and reviews, it is not surprising that Monroe became virtually a household word for sex. It is, for obvious reasons, harder to marshall the evidence for this. I recall it myself and many people I have spoken to remember it too. A couple of quotations may bring it to life.

The first, in the sociological study **Coal Is Our Life** by Norman Dennis, Fernando Henriques and Clifford Slaughter, records the impact of Monroe's appearance in **Niagara** (1953) on a group of miners and their wives in the north-east of England. This is particularly interesting. So much is Monroe part of the coinage of everyday speech, she can be used to exemplify quite different ways of thinking and feeling about sex:

> *In the bookie's office or at the pit they made jokes about the suggestiveness of Miss Monroe, about her possible effect on certain persons present, and about her nickname, 'The Body'. Indeed any man seemed to gain something in stature and recognition if he could contribute some lewd remark to the conversation. On the other hand, in private conversation with a stranger the same*

men would suggest that the film was at best rather silly, and at worst on the verge of disgusting. Finally, the men's comments in the presence of women were entirely different. In a group of married couples who all knew each other well, the women said that they thought Miss Monroe silly and her characteristics overdone; the men said that they liked the thought of a night in bed with her. The more forward of the women soon showed up their husbands by coming back with some remark as 'You wouldn't be so much bloody good to her anyway!' and the man would feed awkward (Dennis *et al,* (1969) p.216).

The second quotation is from Marilyn French's novel, **The Women's Room.** Much of this book is set in the fifties, among a group of newly-weds on a suburban estate. In one section, the narrator (who is also one of the characters) discusses their feelings about sex. This is revealing not only for the inevitability of the Monroe reference, but also for the way it touches upon aspects of sexuality that I'll be dealing with in the rest of this chapter.*

*French is writing retrospectively about the fifities; so am I. The fact that our emphases are similar (besides, I learnt a lot from her novel) should alert us to the possibility that this may be the post-sixties way of constructing the fifties. We should always be aware of the way in which we make over the past in the concerns of the present; but there is a reductio ad absurdum whereby any investigation of the past is held to be *only* a reflection of the present. The relation is more dynamic. French (and I) may be emphasising aspects of the fifties out of post-sixties interests, but that doesn't mean that what is being emphasised was not also a fact of the fifties.

Sex was for most of the men and all of the women a disappointment they never mentioned. Sex, after all, was **THE thing that came naturally,** *and if it didn't – if it wasn't for them worth anywhere near all the furtiveness and dirty jokes and* **pin-up calendars and 'men's' magazines,** *all the shock and renunciation of hundreds of heroines in hundreds of books – why then it was they who were inadequate . . . Probably because most people have an extremely limited sexual experience, it is easy for them, when things are wrong, to place the blame on their partner. It would be different if, instead of graying Theresa with her sagging breasts, her womb hanging low from having held six children, Don were in bed with – Marilyn Monroe, say.* (French, 1978, pp.106-7, my emphasis).

As **The Women's Room** makes clear, sex was seen as perhaps the most important thing in life in fifties America. Certain publishing events suggest this: the two Kinsey reports (on men, 1948; on women, 1953), the first issues of **Confidential** in 1951 and **Playboy** in 1953, both to gain very rapidly in circulation; best-selling novels such as **From Here To Eternity** (1951), **A House is Not a Home** (1953), **Not As a Stranger** (1955), **Peyton Place** (1956), **Strangers When We Meet** (1953), **A Summer Place** (1958), **The Chapman Report** (1960), **Return to Peyton Place** (1961), not to mention the thrillers of Mickey Spillane. Betty Friedan in **The**

Feminine Mystique quotes a survey by Albert Ellis, published as **The Folklore of Sex** in 1961, which shows that 'In American media there were more than 2½ times as many references to sex in 1960 as in 1950' (Friedan, 1963, p.229), and she considers that 'From 1950 to 1960 the interest of men in the details of intercourse paled before the avidity of women – both as depicted in these media, and as its audience' (Ibid: p.230). Nor is this just a question of quantity; rather it seems like a high point of the trend that Michel Foucault has discussed in **The History of Sexuality** as emerging in the seventeenth century, whereby sexuality is designated as the aspect of human existence where we may learn the truth about ourselves. This often takes the form of digging below the surface, on the assumption that what is below must necessarily be more true and must also be what causes the surface to take the form it does. This is equally the model with the psychoanalytical enquiry into the unconscious (peel back the Ego to the truth of the Id), the best-selling novel formula of 'taking the lid off the suburbs' (**Peyton Place** 'tears down brick, stucco, and tar-paper to give intimate revealing glimpses of the inhabitants within', said the **Sunday Dispatch**), or in the endless raking over the past of a star, like Monroe, to find the truth about her personality. And the below-surface that they all tend to come up with in the fifties is sex.

The assumption that sex matters so much is granted even by writers who attacked the directions that they saw sexuality taking. Howard Whitman in his book **The Sex Age,** published in 1962, declares in his foreword:

Of all areas, sex is perhaps the most personal. But it is also a reflection of all of life and of the whole of a culture.

He quotes a Midwest minister as saying:

When men and women come to me with their problems, nine times out of ten as soon as we scratch the surface we find that sex is involved'. (Whitman, 1963, p.3).

Whitman's message is a familiar enough anti-promiscuity, anti-sexual variety, anti-pornography package, but its starting point is that sex is the key to life. For this reason he is anti the wrong kind of sex, but very far from being anti-sex altogether. He quotes H.G. Wells on his title page –

The future of sex is the center of the whole problem of the human future.

Hard to get a clearer declaration than that of how much sex was held to matter in the fifties.

Probably the most lucid interpretation of the fifties' discourses on sexuality remains Betty Friedan's **The Feminine Mystique,** first published in 1963 and clearly a major influence on everything since written about the fifties. Friedan suggests that sexuality at that time became constructed as the 'answer' to any of the dissatisfactions or distress that might be voiced by women as a result of living under 'the feminine mystique', or what Friedan also calls 'the problem that has no name'. Time and again in interviewing women, she would find that they would 'give me an explicitly sexual answer to a question that was not sexual at all' (Friedan, p.226), and she argues that women in America 'are putting into the sexual search all their frustrated needs for self-realisation' (p.289). Similarly, in her survey of some films of the fifties, **On the Verge of Revolt,** Brandon French argues that the films 'reveal how sex and love were often misused to obscure or resolve deeper sources of female (and male) dissatisfaction' (B. French, 1978, p.xxii). If in Foucault's account sexuality is seen as a source of knowledge about human existence, Friedan and French show how that knowledge is also offered as the solution to the problems of human existence. All argue that sexuality, both as knowledge and solution, is also the means by which men and women are designated a place in society, and are kept in their place.

In line with these wider trends in society, sexuality was becoming increasingly important in films. One of the cinema's strategies in the face of the increasingly privatised forms of leisure (not only television, but reading, do-it-yourself, home-based sports, entertaining at home, and so on) was to provide the kind of fare that was not deemed suitable for home consumption – hence the fall of the family film and the rise of 'adult' cinema. Though the huge increase in widely available pornography does not come until later, even mainstream cinema became gradually more 'daring' and 'explicit' in its treatment of sex. Taboos were broken, not only in underground cinema and the rather anti-sex 'hygiene pictures' of the period, but in big Hollywood productions too. Monroe was herself a taboo breaker, from riding the scandal of the nude **Golden Dreams** calendar to showing her nipples in her last photo session with Bert Stern and doing a nude bathing scene in the unfinished **Something's Got To Give,** unheard of for a major motion picture star. Perhaps the most telling manifestations of this more explicit concern, and anxiety, about sexuality are in the characteristic comedies, romances and musicals of the period, which no longer define the problems of hero and heroine in terms of love and understanding, but starkly in terms of virginity – will she, won't she? should I, shouldn't I? As Howard Whitman (1962, p.183) puts it in his 'Why Virginity?' chapter, 'The question is *how far to go*'. What concerns these films is what 'going far' means, because the act of sex is seen as the way to understand the worth and nature of the most privileged of human relationships, the heterosexual couple.

Monroe's image spoke to and articulated the particular ways that sexuality was thought and felt about in the period. This thought and feeling can be organised around two discourses, that of the 'playboy', crystallised by **Playboy** magazine but by no means confined to it; and that of the 'question' of female sexuality itself, at the clinical level revolving around notions of the vaginal orgasm but in popular culture centring more upon the particular, and particularly mysterious or mystifying, nature of female sexual response. These two discourses draw into them many others, and are united by the notion of 'desirability' as the female sexual characteristic that meets the needs of the playboy discourse. Monroe embodies and to a degree authenticates these discourses, but there is also a sense in which she begins to act out the drama of the difficulty of embodying them.

Playboy

It was in 1953 that Monroe was first voted top female box-office star by American film distributors. She was a centre of attraction, in films, promotion and publicity. The first three films in which she had the starring role were released (**Niagara, Gentlemen Prefer Blondes, How To Marry a Millionaire**); she appeared on the cover of **Look** magazine; she walked off the set of **The Girl in Pink Tights;** and, in 1954, she married Joe di Maggio in January and visited the troops in Korea in February. The year 1953 was a time when the most directly sexual of stars was also the star of the moment, and it was also a year of extraordinarily compelling significance in the history of sexuality. In August, the Kinsey report on women was published, with the most massive press reception ever accorded a scientific treatise, and in December the first issue of **Playboy** appeared. The very publication of a sex report on women, and with an attendant publicity far in excess of that surrounding the male report in 1948, focused the 'question' of female sexuality, even if the way in which this question continued to be viewed was actually at considerable variance to Kinsey's

'Golden Dreams' (photo by Tom Kelley)

findings. I'll come back to this, and the relation between Monroe and the question of female sexuality. The **Playboy** connection is more direct.

Monroe was on the cover of the first edition of **Playboy** and inside her **Golden Dreams** nude calendar photo was the magazine's first centrefold. When Molly Haskell observes that Monroe was 'the living embodiment' of an image of woman 'immortalised in **Esquire** and **Playboy'** (Haskell, 1974, p.255)* this is no mere suggestive link between Monroe and **Playboy** – the two are identified with each other from that first cover and centrefold. (Compare also Thomas B. Harris' (1957) discussion of 20th Century-Fox's conscious promotion of her as 'the ideal playmate'.) **Playboy** elaborates the discourse through its developing house style, in copy and photography, and, eventually, in its 'philosophy'. As a star, Monroe legitimates and authenticates this, not just by being in the magazine – though there can be no question of the boost she gave it – but by enacting, as no one else was doing at the time, the particular definitions of sexuality which **Playboy** was proselytising.

*Haskell compares Monroe to the *cartoon* image in these magazines of the showgirl with the old man, but the connection is also with the photographic centrefold image.

Some idea of the double impact, of Monroe on **Playboy,**

and **Playboy** on Monroe, can be gained by considering that **Golden Dreams** centrefold. It was already an object of scandalous interest. The photo had been taken in 1948 by Tom Kelley, and been used for several different calendars, one of hundreds of such images. However, in March 1952, the fact that this image was of an important new Hollywood star became a major news story. At this point, despite the wide circulation of the calendars, relatively few people had actually seen the photo except in small, black and white reproductions accompanying the news item about it. Its scandalousness made it still, in December 1953, an object of much interest, and printing it in a full colour two-page spread in the first issue of a new magazine was a marketing coup.

What is important here is the nature of the scandal **Playboy** so unerringly turned to. On the one hand, there was the fact of a Hollywood star doing a pin-up like **Golden Dreams;** and on the other, there was the widely reported reaction of Monroe when the scandal broke.

The early pin-ups of Monroe belong not to the highly wrought glamour traditions of Hollywood, associated with photographers such as Ruth Harriet Louise and George Hurrell; they belong rather to a much simpler and probably far more common tradition, both in style and choice of model. The style is generally head-on, using high-key lighting, few props and vague backdrops; the model is always young, generally white, the 'healthy, American, cheerleader type' (Hess, 1972, p.227), and not individualised. The key icon of this tradition, certainly in the forties and fifties, is the one-piece bathing costume, whose rigours make all bodies conform to a certain notion of streamlined femininity. But **Golden Dreams** is not like these pin-ups or the Hollywood glamour type; it belongs rather to a tradition known as 'art photography' (since it was ostensibly sold to artists, whose responses to naked women were supposedly less coarse than other men's). In this tradition the model is invariably nude and, though the lighting and camera position are often quite straightforward, the model is usually required to pose in wilfully bizarre positions that run counter to most established notions of classical grace and line. Clearly few were fooled by the art label for this unpleasantly dehumanising tradition of photography – and it was indeed a disreputable form, associated, quite correctly, with the dirty talk of men's locker rooms and toilets.

The scandal was that a Hollywood star had become associated with this tradition, but Monroe's reported reaction took the sting out of the scandal and made the photo just the one **Playboy** needed for its 'new' ideas about sex. In an interview with

Marilyn Monroe in one-piece bathing costume: 'streamlined femininity'

Typical 'art' photography of the fifties: (l.) Photo by Andre de Dienes (r.) Photo from **Sprite** (Los Angeles, undated)

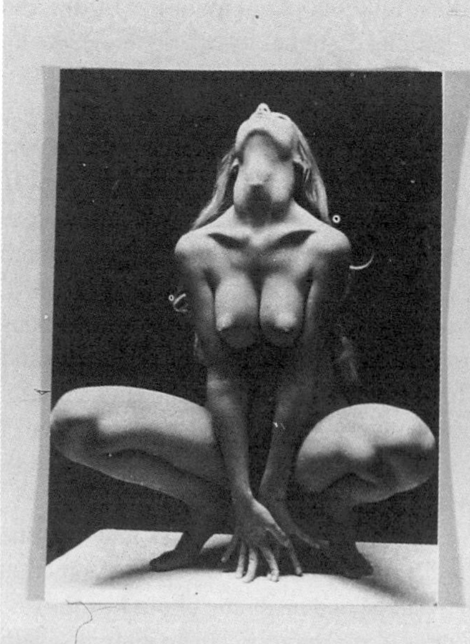

Aline Mosby, Monroe said that she'd done the photo because she needed the money, that Kelley's wife was present at the time, and besides, 'I'm not ashamed of it. I've done nothing wrong' (quoted in Zolotow, 1961, p.105). This sense of guiltlessness is picked up by **Time** magazine, whose wording is, as we'll see, significant – 'Marilyn believes in doing what comes naturally' (**Time,** 11.8.52). They also quote her reply when asked what she had on when the photo was taken – 'I had the radio on.' A classic dumb blonde one-liner, it implies a refusal or inability to answer the question at the level of prurience at which it was asked – indeed, it suggests an innocence of prurience altogether.

Guiltless, natural, not prurient – these were precisely part of the attitude towards sexuality that **Playboy** was pushing. **Playboy's** 'philosophy' – not formally articulated as such until 1962, but clearly developing through the magazine in the fifties – combined two reigning ideas of the twentieth century concerning sexuality. The first is what Michel Foucault has called 'the repressive hypothesis', namely, the idea that sexuality has 'been rigorously subjugated . . . during the age of the hypocritical, bustling and responsible bourgeoisie' (Foucault, 1980, p.8). The second has been termed by John Gagnon and John Simon (1974) a 'drive reduction model' of sexuality, positing the sex drive as 'a basic biological mandate' seeking 'expression' or 'release'. It is common enough to see this 'biological mandate' as a fierce and disruptive drive which really needs repression, but **Playboy's** view of it was benign – only repression itself turns the sex drive malignant, and left to its own devices it will bring nothing but beauty and happiness:

There are a great many well-meaning members of our own society who sincerely believe that we would have a happier, healthier civilization if there were less emphasis upon sex in it. These people are ignorant of the most fundamental facts on the subject. What is clearly needed is a **greater** *emphasis upon sex, not the opposite. Provided, of course, we really do want a healthy, heterosexual society.*

A society may offer negative, suppressive, perverted concepts of sex, relating sex to sin, sickness, shame and guilt; or, hopefully, it may offer a positive, permissive, natural view, where sex is related to happiness, to beauty, to health and to feelings of pleasure and fulfilment.

Sex exists with and without love and in both forms it does far more good than harm. The attempts at its suppression, however, are almost universally harmful, both to the individuals involved and to society as a whole.

The force of these ideas for the mid-twentieth century, and especially American, common-sense thought lies in their appeal to the idea of naturalness, the idea that you can justify any attitude or course of action by asserting it to be in accord with what people would really be like if they lived in a state of nature. Sexuality is peculiarly amenable to this kind of argument, since it is at first glance (our habitual first glance, anyhow) so 'biological', so rooted in the flesh. Monroe, so much set up in terms of sexuality, also seemed to personify naturalness. Her perceived naturalness not only guaranteed the truth of her sexuality, in much the same way as imputed qualities of sincerity and authenticity, spontaneity and openness, guarantee the personality of other stars; it was also to define and justify that sexuality, exactly in line with the **Playboy** discourse.

The assertion of Monroe's naturalness in relation to sexuality has been made so often that I do not need to establish it at length. At the time, critics and observers referred to it constantly, and retrospectively many of the people involved with her have ascribed it to her. Jayne Mansfield, when asked to appear nude at a nudist colony in Rio de Janeiro, refused, reportedly saying:

It's too bad I'm not Marilyn Monroe. She's a naturalist. But I would not feel right.

Though always thought of as an also-ran Monroe, Mansfield clearly recognised the particular ingredient in Monroe's image that she herself did not have. The pose and expression of each in otherwise similar star bath tub photos clearly captures the difference. Immediately following Monroe's death, Diana Trilling (1963, p.236) wrote an article on her which is in many ways emblematic of this widely held view of Monroe:

None but Marilyn Monroe could suggest such a purity of sexual delight.

And Monroe herself said in her last interview:

I think that sexuality is only attractive when it is natural and spontaneous.

Marilyn Monroe in bath-tub (photo Sam Shaw)

Jayne Mansfield in bath-tub

We have to tread carefully here, since ambiguities crowd in: for instance, most of those who ascribe 'natural' sexuality to her are in fact describing their *response* to her. There is also present, in so much of the writing, an endless raking over of the possible perversities (= unnatural nature) of Monroe's 'real' sexuality. What we need to keep in focus is the degree to which the Monroe image clearly offered itself to be read in terms of (benign) naturalness and with the impact of being something new. A promotion photo, a gag and a couple of films will serve to illustrate this.

In 1950, when Monroe had been signed to a seven-year contract with 20th Century-Fox, she was photographed for the studio with a group of other contract players by Philippe Halsman. It is not only with hindsight, because she is the only one we now recognise, that Monroe stands out. We may ascribe to Halsman the fact that Monroe is placed at the front and in the centre and looks straight to camera (rather than in the various off-screen or self-absorbed directions of the other players) and thus seems to make a direct contact with the viewer's eyes. We

20th Century-Fox contract players in 1950 (photo Philippe Halsman)

could go on to ascribe the very simple, relaxed pose to Halsman, the tousled, apparently naturally falling, uncoiffed hair to an expert hair stylist, the unfussy blouse to the wardrobe department and so on. But of course we have no way of knowing who made such decisions, Monroe or someone else; the evidence in the biographies suggests that even if others did make the decisions, it was because they had already ascribed naturalness to her in their minds; and, most important, it is unlikely that anyone seeing this photo in 1950 would have sought to identify those responsible for constructing her in an image of naturalness. Indeed, what is striking about the photo is the contrast between the very obviously contrived poses of the other players, though each a very recognisable female stereotype of the period, and the apparently artless look of Monroe that makes the others seem constructed but her seem just natural. Many other Monroe pin-ups from around this time have a similar quality, and the contrast between Monroe and the others in this one photo encapsulates the more general contrast that was beginning to be apparent between her pin-ups and the other available cheesecake.

This may seem like a laborious treatment of the question, but it is I think important to state it as precisely as possible. Monroe did appear natural in her sexiness and with an originality that necessarily had an impact among the stream of conventionally pretty starlets and pin-ups that the studios continually produced. It may only have been appearance, but we are dealing in appearances and what they are taken to mean. To put it another way, it seems reasonable to suggest that the quality of naturalness, so crucial for **Playboy** in its first centrefold, would probably not have been conveyed by any of the other players in the Halsman photo, nor indeed by any of the Monroe lookalikes such as Jayne Mansfield or Mamie Van Doren – and not merely because the **Golden Dreams** centrefold expressed it, but rather because the Monroe image of naturalness was, by the time the calendar photo was reprinted in 1953, already powerful enough to make the effect one of 'naturalness'.

There had been female star images that suggested naturalness before, but usually in a context that said very little about sexuality. The woman sitting on the floor next to Monroe in the Halsman photo has that asexual (or perhaps just covertly sexual) naturalness of which June Allyson was one of the most charming exponents. Monroe combined naturalness *and* overt sexuality, notably in a series of gags that became known as Monroeisms. Though in form typical of the dumb blonde tradition to which she in part belongs (cf. Dyer, 1979), they are different in being nearly always to do with sex. One of the most striking is one delivered to the troops in Korea in February 1954:

I don't know why you boys are always getting excited about sweater girls. Take away their sweaters and what have they got?

Though overtly referring to other women stars, she effectively refers to herself, her own body and perhaps even her own breasts so recently exposed in **Playboy.**

Though a clever gag, it is also, in context, dumb, because Marilyn Monroe is a dumb blonde. The dumbness of the dumb blonde is by tradition natural, because it means that she is not touched by the rationality of the world. She is also untouched by the corruption of the world; a figure out of Rousseau, but some way from his conceptions of the essential nature of the human being before civilisation gets to her or him. The dumb blonde's ignorance of the world is brainless, seldom the superior wisdom of Rousseau's 'natural' women; and her innocence is above all a sexual innocence, a lack of knowledge about sexuality. She is a figure in comedy, because she is also always extraordinarily and devastatingly sexually attractive – the comedy resides either in the way her irresistible attractions get men tied up in her irrationality or else in the contrast between her sexual innocence and her sexual impact. The most interesting play on these comic possibilities comes about when ambiguities are acknowledged – maybe (as with Judy Holliday in **It Should Happen To You**) the dumb blonde's irrationality is the wisdom of the Holy Fool; maybe (as with Carol Channing in the stage version of **Gentlemen Prefer Blondes**) she is using the dumb blonde image to manipulate men. But Monroe's image does not really follow either of these directions – rather she fundamentally alters the dumb blonde comic equation. Rationality hardly comes up as a question in the comedy of her films at all, it is sexual innocence that's the core of the gags – but it is no longer a contrast between sexuality and innocence, since with Monroe sexuality *is* innocent. So the sweater girl gag is not funny because the blonde is being dirty about herself without knowing it, but because it is a play on words that cheerfully acknowledges her sexual impact. Monroe knows about sexuality, but she doesn't know about guilt and innocence – she welcomes sex as natural.

Several Monroe films play on this innocent/natural attitude to sexuality. **The Misfits** is a sustained equation of

Marilyn Monroe and Cary Grant in **Monkey Business** (1952)

Marilyn Monroe pin-up: 'she throws her head back . . . she opens her mouth' (photo Michael Conway and Mark Ricci)

Monroe/Roslyn with nature, including in this an easy attitude towards sex. **Monkey Business** and **The Prince and the Showgirl** are in some ways more interesting. The plot of **Monkey Business** concerns a rejuvenation drug. When middle-aged people take it, they become young again, meaning both uncivilised (like the monkeys in the research laboratory) and sexy. For the Cary Grant character (Barnaby Fulton), this is realised through his change in response to the Monroe character (Miss Laurel). When earlier she shows him her leg (in order to display her new stockings), he is merely embarrassed; but after he has inadvertently taken the drug he embarks on a free-wheeling, spontaneous, youthful (= natural) escapade with Miss Laurel – but whereas the joke is that this is him letting his hair down, she is clearly just getting into it because that's the way she is normally. Tearing along the highway in a sports car he has impulsively (= naturally) bought, she throws her head back, her hair flutters in the breeze, she opens her mouth and giggle-laughs. It is *the* Monroe image, here exactly placed to mean the natural enjoyment of sensation. That this enjoyment includes sexuality is made clear elsewhere in the sequence.

The Prince and the Showgirl makes a Jamesian equation between Monroe as American, child-like, emotional, direct, and Olivier as European, adult, rational, sophisticated. When Olivier (the Regent) makes a flowery, melancholy speech about being 'a sleeping prince that needs the kiss of a beautiful young maiden to bring him back to life', Monroe (Elsie) says, 'You mean you want me to kiss you'; to which he wearily replies, 'You're so literal'. Elsie/Monroe here straightforwardly accepts the sexual, without coyness, embarrassment or sniggering. But the film is actually rather incoherent in relation to Elsie's innocence. At the beginning, Elsie/Monroe is repeatedly given lines that indicate she doesn't know why the Regent has invited her to the embassy – 'Tough question that, all right', says her flat mate, in precisely the kind of dry, knowing, wise-cracking voice that Monroe would never use. However, when she sees the dinner for two brought in, Elsie/Monroe says that she knows 'every rule' in the sex game and starts to walk out. Such contradictions, with Elsie at one moment ignorant of sexual game-playing and at the next more than conversant with it, run through the film, so that most often it misses that combination of knowledge of sexuality without loss of innocence which is one of the keys to Monroe's image.

Naturalness, which Monroe so vividly embodied and thereby guaranteed, was elaborated in **Playboy** above all at the level of its 'philosophy', its overt and proclaimed Weltanschauung. At this level, it's an attitude that sees itself as socially progressive, taboo-breaking. The feeling it conveys is exactly that noted by Michel Foucault (1980, pp.6-7) as characteristic of those committed to 'the repression hypothesis':

We are conscious of defying established power, our tone of voice shows that we know we are being subversive, and we ardently conjure away the present and appeal to the future, whose day will be hastened by the contribution we believe we are making.

Foucault's irony stems from his proposition that we should not think so much in terms of sexuality being repressed, but rather in terms of its form being constructed, and with an ever renewed insistence, as an instrument of power. Similarly the feminist critique of the **Playboy** discourse points out that what it is concerned with is new definitions of male power within sexuality. Yet **Playboy's** own retrospective view of itself may not be so wide of the mark either, in suggesting how its emergence felt to many people at the time:

Playboy *came out of aspects of the same energy that created the beat crowd, the first rock-'n'-rollers, Holden Caulfield, James Dean,* **Mad** *magazine – and anything else that was interesting by virtue of not eating the prevailing bullshit and being therefore slightly dangerous (***Playboy,** January 1979).

Playboy was not only its declared philosophy, it was the whole package, and especially its playmate centrefolds. If overtly **Playboy** wanted to overthrow a hide-bound society, much of what it did in its pages seems an attempt to integrate its sexual freedom into suburban and white-collar life – itself pretty well taken as the norm in fifties' iconography (hence the popular success of the symptomatically titled novel, **The Man in the Grey Flannel Suit,** 1955). **Playboy's** greatest success was to get itself sold in the most ordinary newsagents and drugstores, taking a sex magazine out of the beneath-the-counter, adult-bookshop category. Its success in doing this resulted partly from attracting name writers and other such strategies, but the centrefolds also played their part. What **Playboy** succeeded in doing was making sex objects everyday.

David Standish, writing in **Playboy's** twenty-fifth anniversary edition, suggests that **Playboy's** aim was to present 'a pin-up as something other than a porno postcard', as, in fact, 'the girl next door' (art photography meets June Allyson = Marilyn Monroe), and he takes the July 1955 centrefold of Janet Pilgrim as the turning point in this project.

At the time, the idea that a 'nice' girl would appear in the four-colour altogether was shocking! . . . Suddenly, here were girls, **a** *girl, Janet Pilgrim, who looked like a good, decent human being and worked in an actual office . . . not some distant, bored bimbo with her clothes off but, perhaps, if God were in a good mood, she might one day be that girl you see on the bus every day who's making your heart melt (*ibid.).

The Seven Year Itch, made in 1954, works off just this fantasy. Monroe is the never-named girl upstairs, the kind of girl who appears in art photo magazines of the kind that Richard Sherman (Tom Ewell) buys, and who just happens to move into the apartment upstairs. It's the **Playboy** dream come true. At the end of the film, as he is leaving to rejoin his wife, he calls after the girl, 'What's your name?' 'Marilyn Monroe,' she jokes back, the film thus signalling that it knows how inextricable are the Monroe and playmate images.

Janet Pilgrim, the July 1955 centrefold, is almost, in the way she is written about, a re-run of Monroe's career. To quote Standish again, she is 'an engaging blonde' (more on blondeness later) 'shown first at work slaving beautifully over her typewriter' (Monroe had played a secretary in **Hometown Story,** 1951; **As Young As You Feel,** 1951 and **Monkey Business,** 1952), 'then sitting two pages later wearing mostly diamonds at a fancy dressing table' (Monroe's biggest number in **Gentlemen Prefer Blondes** (1953) was *Diamonds Are a Girl's Best Friend,* so big that it is also used for her appearance at the fashion show in **How to Marry a Millionaire,** 1953). Pilgrim/Monroe normalises sex appeal (in the secretary image) whilst still associating it with something to be possessed, like a mistress bought with diamonds.

Let's pause, though, on the secretary image. When secretarial work first developed towards the end of the nineteenth century, it was a prestigious job for women that could be looked on as both interesting work and a source of advancement; by the fifties, it had become a routine job. This is reflected in fiction aimed at women, where, according to Donald R. Maskosky's (1966, p.38) survey of women's magazine stories, in the fifties 'the image of the secretary . . . is often of a competent employee who should, however, not expect advancement. Her hope for advancement lies instead in matrimony'. In films of the fifties not aimed specifically at women, the dynamic of 'advancement' does not appear at all. Secretarial work is almost totally unseen; secretaries are there for the men in the office to look at (compare Jo Spence's (1978/79) discussion in her article 'What Do People

Do All Day?'). There is a scene in **As Young As You Feel** where two policemen come into an outer office where Monroe (Harriet) works. We see her working, standing looking at herself in front of the mirror; she shows the policemen into her boss's room and then returns to her desk to work – combing her hair. As the films construct the function of the secretary, this *is* her work, preparing herself to be looked at. In **Monkey Business** the fact that she can't do secretarial work is underlined, to be dismissed with the gag – as Cary Grant and Charles Coburn's eyes follow her out of the door – 'Well, anyone can type'. Later Ginger Rogers (Edwina) refers to her jealously as 'that little pin-up', once again establishing the playboy conflation of secretary and sex object.

Such a woman is there for men. This is the nub of the playboy discourse; its unstated assumption is that 'sex is for the man', in the words of the working-class married couples interviewed by Lee Rainwater in his 1960 study, **And The Poor Get Children.** Women are set up as the embodiment of sexuality itself. As Hollis Alpert (1956, p.38) put it at the time – and presumably without any intended feminist irony:

Hollywood has given [audiences] the Hollywood Siren – the woman who simply by existing, or at most sprawling on a rug or sauntering up a street – is supposed to imply all the vigorous, kaleidoscopic possibilities of human sexuality.

Women are to *be* sexuality, yet this really means as a vehicle *for* male sexuality. Monroe refers to her own sexualness – her breasts in the sweater girl gag, or her buttocks in the line near the beginning of **The Seven Year Itch,** 'My fan is caught in the door' – but read through the eyes of the playboy discourse, she is not referring to a body she experiences but rather to a body that is experienced by others, that is, men. By embodying the desired sexual playmate she, a woman, becomes the vehicle for securing a male sexuality free of guilt.

The sexuality implied by the playboy discourse and Monroe (in so far as the two are to be equated) can seem like something out of Eden, and the idea echoes around the fifties and early sixties. Maurice Zolotow, in his prurient biography **Marilyn Monroe the Tragic Venus** (1961, p.94), cuts through any sense of the complexity of human sexual response with

There are few pleasures as immediate and uncomplicated as the sight of a comely naked girl.

When Monroe came to England to film **The Prince and the Showgirl** in 1956, the **Evening News** wrote, 'She really is as luscious as strawberries and cream', and in his 1973 **Marilyn,** very much written out of a fifties sensibility, Norman Mailer (1973, p.15) endlessly reaches for similar imagery:

Marilyn suggested sex might be difficult and dangerous with others, but ice cream with her.

. . . on the screen like a sweet peach bursting before one's eyes . . .

. . . so curvaceous and yet without menace . . .

Mailer here makes explicit what others had sensed without knowing, that the Monroe playmate is an escape from the threat posed by female sexuality. For, as the **Readers' Digest** pointed out in 1957, 'What Every Husband Needs' is, simply, 'good sex uncomplicated by the worry of satisfying his woman' (quoted by Miller and Nowak, pp.157–8).

Desirability

Monroe not only provided the vehicle for expressing the playboy project of 'liberating' sexuality, she was also the epitome of what was desirable in a playmate. 'Desirability' is the quality that women in the fifties were urged to attain in order to make men (and thereby themselves) happy. In 1953 Lelord Kordel, for instance, declared in **Coronet:**

The smart woman will keep herself desirable. It is her duty to herself to be feminine and desirable at all times in the eyes of the opposite sex' (quoted by Miller and Nowak, 1977, p.157).

'In the *eyes* of . . .'; the visual reference is striking despite being also so commonplace.

Monroe conforms to, and is part of the construction of, what constitutes desirability in women. This is a set of implied character traits, but before it is that it is also a social position, for the desirable woman is a white woman. The typical playmate is white, and most often blonde; and, of course, so is Monroe. Monroe could have been some sort of star had she been dark, but not the ultimate embodiment of the desirable woman.

To be the ideal Monroe had to be white, and not just

white but blonde, the most unambiguously white you can get. (She was not a natural blonde; she started dyeing her hair in 1947.) This race element conflates with sexuality in (at least) two ways. First, the white woman is offered as the most highly prized possession of the white man, and the envy of all other races. Imperialist and Southern popular culture abounds in imagery playing on this theme, and this has been the major source of all race images in the twentieth century. Thus there is the notion of the universally desired 'white Goddess' (offered at the level of intellectual discourse, in 'anthropological' works such as Robert Graves' **The White Goddess,** as a general feature of all human cultures), and explicitly adumbrated in Rider Haggard's **She** and its several film versions. There is the rape motif exploited in **The Birth of a Nation** and countless films and novels before and since; and there is the most obvious playing out of this in **King Kong,** with the jungle creature ascending the pinnacle of the Western world caressingly clutching a white woman. (In the re-make Jessica Lange affects a Monroe accent for the part.)

Blondeness, especially platinum (peroxide) blondeness, is the ultimate sign of whiteness. Blonde hair is frequently associated with wealth, either in the choice of the term platinum or in pin-ups where the hair colour is visually rhymed with a silver or gold dress and with jewellery. (We might remember too the

Marilyn Monroe pin-up: 'the ultimate sign of whiteness'

title of Monroe's nude calendar pose, **GOLDEN Dreams.**) And blondeness is racially unambiguous. It keeps the white woman distinct from the black, brown or yellow, and at the same time it assures the viewer that the woman is the genuine article. The hysteria surrounding ambiguity on this point is astonishing. **Birth of a Nation** comes close to suggesting that congressman Stevens' mulatto housekeeper was a major cause of the civil war; the fact of being half-caste makes Julie into a tragic character in **Show Boat;** and the *thought* that she might be half-caste sends Elizabeth Taylor mad in **Raintree County.** (All these films, one might add, are based on best-selling popular novels.) The film career of Lena Horne is also instructive: as a very light-skinned black woman, she was unplaceable except as the ultimate temptress in an all-black musical, **Cabin in the Sky,** where the guarantee of her beauty resides in the very fact of being so light. Otherwise she could not really be given a role in a film featuring whites, because her very lightness might make her an object of desire, thus confusing the racial hierarchy of desirability.

The white woman is not only the most prized possession of white patriarchy, she is also part of the symbolism of sexuality itself. Christianity associates sin with darkness and sexuality, virtue with light and chastity. With the denial of female sexuality in the late nineteenth and early twentieth century (except as by definition a problem), sexuality also becomes associated with masculinity. Men are then seen as split between their baser, sexual, 'black' side and their good, spiritual side which is specifically redeemed in Victorian imagery by the chastity of woman. Thus the extreme figures in this conflation of race and gender stereotypes are the black stud/rapist and the white maiden. By the fifties, such extremes were less current, nor did they necessarily carry with them the strict moral associations of sexual = bad, non-sexual = good; but the associations of darkness with the drives model of masculine sexuality and of fairness with female desirability remained strong. The central sexual/love relationship in **Peyton Place** (the original novel), between Connie Mackenzie and Michael Kyros, works very much through such an opposition. Connie's character is established through the admiration of her daughter's friend Selena (dark-haired, lower-class, soon deflowered): Selena wishes that she too had 'a wonderful blonde mother, and a pink and white bedroom of her own' like Alison, Connie's daughter (Metallious, 1957, p.39). As for Michael, the narrator explicitly defines him as 'a handsome man, in a dark-skinned, black-haired, obviously sexual way' (ibid., p.103). The townspeople refer to the couple as 'that big, black

Greek' and 'a well-built blonde' (ibid., p.135). Their relationship is sealed when he makes love to her 'brutally, torturously' (p.135), that is, when this desirable woman is taken by his male drive. Thus in the elaboration of light and dark imagery, the blonde woman comes to represent not only the most desired of women but also the most womanly of women.

Monroe's blondeness is remarked upon often enough in films, but only the first saloon scene in **Bus Stop** seems to make something of it. Beau storms in and at once sees Cherie on stage, the angel that he has said he is looking for. His words emphasise her whiteness – 'Look at her gleaming there so pale and white.' He finds in her the projection of his desires, and the song she sings might be her acknowledgement of this – *'That old* **black** *magic that* **you** *weave so well'*.

Besides blondeness, Monroe also had, or seemed to have, several personality traits that together sum up female desirability in the fifties. She looks like she's no trouble, she is vulnerable, and she appears to offer herself to the viewer, to be available. She embodies what, as quoted at the end of the last section, 'Every Husband Needs' in a wife, namely, good sex uncomplicated by worry about satisfying her. Once again, Norman Mailer articulates this way of reading Monroe – 'difficult and dangerous with the others, but ice cream with her'. Monroe, an image so overdetermined in terms of sexuality, is nevertheless not an image of the danger of sex: she is not the femme fatale of film noir and of other such hypererotic star images as Clara Bow, Marlene Dietrich, Jean Harlow and even Greta Garbo, all of whom in some measure speak trouble for the men in their films. Round about the time Monroe was becoming a major star, 20th Century-Fox did put her in two such roles – as a psychotic baby-sitter in **Don't Bother to Knock** in 1952 and as an adultress in **Niagara** in 1953. Though commercially successful (almost any film with her in it would have been at this point), they were clearly not right for her,* as the reviews for **Niagara,** especially, register. Denis Myers in an article on Monroe in **Picturegoer** (9.5.53) clearly sees how her appeal is separate from any sense of her being dangerously sexual:

*This is not meant as a judgement of her acting capacities. Many people, both at the time and subsequently, consider her performance in **Don't Bother to Knock,** in particular, to be extremely 'good' (this is not the place to debate what criteria are in play here). What's at issue is not whether Monroe, as actress, could play 'dangerous' women, but whether her image allowed these roles to make sense if she played them. I'm arguing that on the whole it did not.

> *In* **Niagara** *she has to convince us that she is desirable. Marilyn does. But – a* **femme fatale?** *We-ell . . .*

Several of the big 20th Century-Fox vehicles seem, at script level, to give her a role with some castrating elements – as a gold-digger in **Gentlemen Prefer Blondes** and **How to Marry a**

Millionaire she sets out to manipulate male sexual response for money, while in **There's No Business Like Show Business** she plays a showgirl who uses Tim's (Donald O'Connor) interest in her to further her career. But she's simply too incompetent, 'dumb' and, to add to it, short-sighted in **How to . . .**, winding up with a bankrupt, and in **There's No Business . . .** the plot makes it clear that she wasn't really two-timing Tim. **Gentlemen . . .** is a more difficult case, but it seems to me that Monroe doesn't play the part as if she is a manipulator. (But see my discussion in **Stars,** pp.147-8 and Pam Cook's different reading in **Star Signs,** pp.81–2.)

In the later roles the disruption that any introduction of a highly sexual (almost the same thing as saying *any*) woman into a male character's life always involves, is defused; indeed it almost becomes the point of the films that Monroe takes the sting out of anything that her sexuality seems likely to stir up. So Richard (Tom Ewell) in **The Seven Year Itch** goes back happily to his wife, Beau (Don Murray) in **Bus Stop** gets his girl (Cherie/Monroe) and goes back to his ranch, Elsie (Monroe) in **The Prince and the Showgirl** reconciles the King (Jeremy Spenser) and his father, the Regent (Olivier), and so on. It's a standard narrative pattern – a state of equilibrium, a disruption and a return to equilibrium through resolution of the disruption; only here the cause of the disruptions (Monroe, just because she *is* sex) and the resolution are embodied in one and the same person/character (Monroe).

If Monroe's desirability has to do with her being no trouble, it also has to do with being vulnerable. Susan Brownmiller (1975, pp.333) in her study of rape, **Against Our Will,** suggests there is 'a deep belief . . . that our attractiveness to men, or sexual desirability, is in direct proportion to our ability to play the victim'. Women live 'the part of the walking wounded' and this is something that 'goes to the very core of our sexuality'. Brownmiller quotes Alfred Hitchcock saying that he looked for 'a certain vulnerability' in his leading ladies, and she points out that the dictionary definition of 'vulnerable' is 'susceptible to being wounded or hurt, or open to attack or assault' (p.334). Thus what made Hitchcock's women stars right was that 'they managed to project the feeling that they could be wounded or "had"'. Brownmiller adds, 'And I think Hitchcock was speaking for most of his profession'. She names Monroe as perhaps 'the most famous and overworked example' of 'the beautiful victim' syndrome (p.335).

Monroe is not generally physically abused in films. She is, rather, taken advantage of or humiliated. Very often this means little more than putting her in situations where she is exposed to the gaze of the male hero, but in two of the films that are also considered her best, **Bus Stop** and **Some Like It Hot,** this goes much further. In **Bus Stop,** she plays Cherie, a show girl who wants to get out of the cheap bar-rooms where she works, to be a success and 'get a little respect'. But even though this was the film set up for her return to Hollywood (after walking out and going to study at the Actors' Studio in New York), and she is undoubtedly the star of it, the project that carries the narrative is not Cherie's, but Beau's (Don Murray). He is looking for his 'angel' and finds her in Cherie/Monroe; the trajectory of the narrative is the defeat of her project in the name of his (getting her to marry him). One of the turning points in the film – and one we are obviously meant to find funny – occurs when Beau, an expert cowboy, lassoes Cherie as she is trying to escape him on a bus. It is not just that the narrative shows her as helpless before the male drive to conquer; the film invites us to delight in her pitiful and hopeless struggling.

Some Like It Hot is even more insidious, for its comedy depends upon plot strategies whereby Monroe/Sugar makes herself defenceless because she thinks she's safe. She is trying to escape men because of all the rotten deals they've dealt her – this is why she's joined an all woman band. Because she is trusting (and because, like any farce, **Some Like It Hot** depends upon characters in the film believing in disguises that are transparent to the audience), the film gets her into situations where she drops her guard; notably in a scene with Joe (Tony Curtis), who's in drag, in the ladies' toilet on the train to Florida. Precisely because she thinks she is in the safety of woman's space, she does not protect herself from him. Before his ogling eyes (and, of course, ours), she lifts her skirt to take a brandy flask out of her garter and titivates her breasts in front of the mirror. Because they are actions a woman would not make in front of a man, Joe/Curtis and the assumed male audience are violating both Monroe and women's space. Moreover, she then sets up the means for further violation. She tells Joe that she wants to marry a rich man who wears glasses, and, armed with this information, he changes his disguise from drag to a short-sighted oil millionaire. In one of the most remembered scenes in the film, on board 'his' yacht, he also pretends to be impotent. Once again, believing she is safe, Sugar/Monroe drops any defence against his sexual harassment, drapes herself over him and kisses him long and languorously. The pleasure we are offered is not just that Marilyn Monroe is giving herself to a man (a potential surrogate for the audience), but that her defences are down, we've got her where we (supposedly) want her.

Monroe's vulnerability is also confirmed by aspects of her off-screen image, which could, indeed, be read as a never-ending series of testimonials to how easily, and frequently, she is hurt. A brief list of the main points that were so often raked over in the publicity surrounding her will suffice to indicate this, always bearing in mind that some of these never happened or are very exaggerated:

born illegitimate to a mother who spent her daughter's childhood in and out of mental hospitals;
fostered by several different couples;
time spent in an orphanage (sometimes presented in Dickensian terms in the biographies, articles and interviews);
indecently assaulted at the age of nine;
an habitual sufferer from menstrual pains;
three unsuccessful marriages;
unable to bear children, having a succession of miscarriages;
a nymphomaniac who was frigid (oh, the categories of fifties' sexual theory!);
a woman so difficult to work with Tony Curtis said kissing her was like kissing Hitler;
a suicide, or murdered, or died of an overdose of the pills she habitually took.

It's a threnody so familiar that all retrospective articles, and references to her, invoke it, and most find quotations from Monroe to do so. Take, for instance, two books on famous people of the twentieth century who have died young. Marianne Sinclair, in **Those Who Died Young,** quotes the poem written by Monroe and published posthumously in **McCalls** in 1962:

Help! Help!
Help! I feel life coming back
When all I want is to die.

Patricia Fox Sheinwold (1980), in **Too Young to Die,** uses another quote:

I always felt insecure and in the way – but most of all I felt scared. I guess I wanted love more than anything in the world.

Thus the image insists that Monroe suffered, and experienced her suffering vividly throughout her life.

The appeal of this biographical vulnerability necessarily involves the power of the reader, but we need to get the emphases right here. Vulnerability may call forth any number of responses, including empathy and protectiveness as well as sadism. It is the way that the Monroe biography is ineluctably associated with sexuality that is significant – not just sexual experience itself, but the inter-relations of sexuality with menstruation, childbirth, marriage, and so on. Monroe's problems are repeatedly related (often using her own words) to the need for love, meaning in the vocabulary of the fifties (hetero)sexual love.

Unthreatening, vulnerable, Monroe always seemed to be available, on offer. At the time, and even more subsequently, many observers saw her career in terms of a series of moments in which she offered herself to the gaze of men – the **Golden Dreams** calendar, **The Seven Year Itch** subway gratings pose, shot before

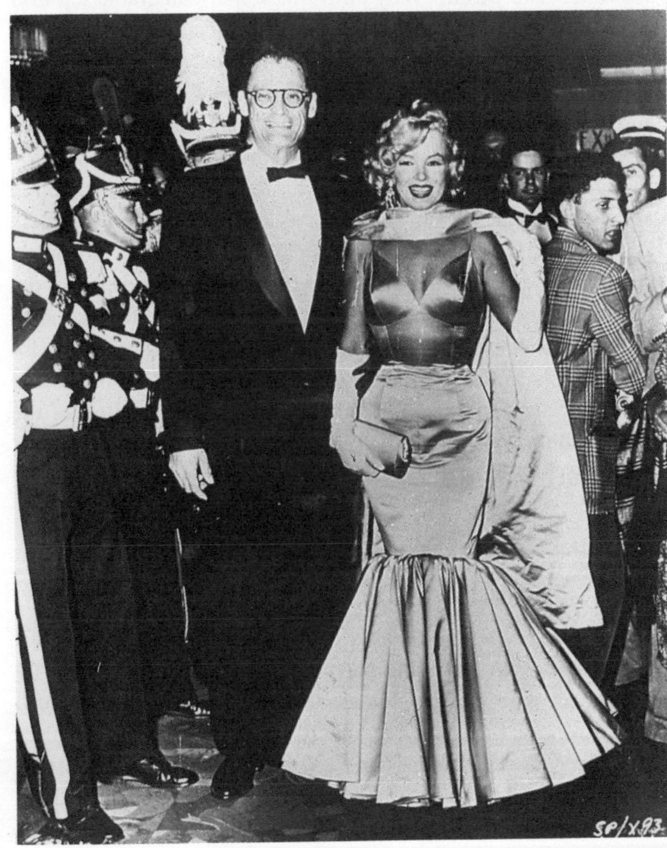

At the première of **The Prince and the Showgirl** with Arthur Miller: 'revealing and fetishistic gowns'

passing crowds in a Manhattan street, her appearances at premières in revealing and fetishistic gowns, her final nude photo session with Bert Stern and nude scene for **Something's Got to Give . . .** All these were taken as done by Monroe, the person, at her own behest. Each one a dramatic news story, they were read not as media manipulation but rather as a star's willing presentation of her sexuality to the world's gaze. Interviews could also be raided for corroborations. Maurice Zolotow (1961) quotes Monroe's words in 1950 to Sonia Wolfson, a publicity woman at 20th Century-Fox, on the subject of the first time she put make-up on:

> *This was the first time in my life I felt loved – no one had ever noticed my face or hair or me before.*

In her last interview, with **Life,** she told of the effect wearing a sweater had had on the boys at school, an effect she revelled in. So many incidents, so many remarks in interviews – if Monroe was a sex object she was not only untroublesome, vulnerable but also seemed to enjoy and promote her own objectification. She was the playboy playmate who wanted to be one.

Wanted to be . . . In the light of the women's movement and its exploration of the formation of human desire, the idea that anyone simply 'wants' to do something, out of a volition untouched by social construction, is untenable. Monroe appeared at a moment when feminism was at its lowest ebb in the twentieth century, and both her career decisions and remarks in interviews could and were read as confirming the male-serving myth of the desirable playmate. But so great an emphasis on her own purported involvement in the production of her sexy image is *also* an emphasis on the will and desire of the person who inhabits and produces the sexy image. It actually raises the question of the person who plays the fantasy, in other terms, the subject who is habitually the object of desire.

Psycho

The image of the desirable playmate, which Monroe so exactly incarnated, is an image of female sexuality for men. Yet so much does it insist on the equation women = sexuality, that it also raises the question, or spectre, of female sexuality for women. Monroe's image is much less clearly articulated in relation to this – part of what makes her desirable, unthreatening, is that her image does not insist on a female sexuality for itself. Yet there is always the possibility of the other, woman-centred reading. On the one hand, she registers aspects of the fifties discourse on the 'question' of female sexuality; and on the other, some of her actions and performances begin to articulate something of the drama and difficulty of embodying the desirable playmate – part of that general, beneath the surface, not yet articulable expression of discontent that Brandon French (1978) takes to indicate that in the fifties women were, in the title of her book, 'on the verge of revolt'.

The most decisive discourse on female sexuality at this time was that stemming from psychoanalysis. There was a massive spread and popularisation of psychoanalytical ideas, both in magazines and fiction and in the developing practices of counselling, family guidance, social work, and so on, to say nothing of the growth of therapy proper. This is clearly marked in the films of the period, above all in the melodramas with their very direct acknowledgement of the unconscious, of repression, of drives, of the sexuality of family relations, of the sexuality of everything. The growth of the 'psychological Western', the knowing gags of the sex comedies and the development of Method acting are further indices of how far psychoanalytic ideas, however garbled, had got under the skin of popular culture. Perhaps one of the most striking, because unexpected, indications of the trend is MGM's sci-fi extravaganza, **Forbidden Planet,** whose plot depends upon the audience being able to grasp the notions of the Id and the Ego. This can be termed the psycho discourse, since Hitchcock, whose films are so marked by explicit and implicit references to psychoanalysis, perfectly caught – in naming **Psycho** – the popularisation of psychoanalysis, such that a whole world of reference could be summed up in a jaunty abbreviation.

In the psycho discourse, female sexuality is always dependent on male sexuality – whether because it is seen purely in terms of a response to male sexuality, as an internalisation of male desire, or as an experience that can only be had in heterosexual intercourse. The first of these male-oriented views of female sexuality takes us back to desirability and the idea of 'readiness'; the second leads to notions of narcissism, while the third raises the question of the nature of female orgasm.

Desirability is what makes a woman attractive to a man; but in the psycho discourse it is also a source of pleasure and satisfaction for the woman herself. It is the concomitant of the playboy drive reduction model of male sexuality, a model deriving from one version of psychoanalytic thought. As Liz Stanley (1977, p.10) puts it in her paper, **'The Problematic Nature of Sexual Meanings',**

> *the active sexual drive requiring expression is seen as a* **male** *drive . . . Female sexuality is seen as a recipient of male sexual drives and not as an active outgoing sexuality in its own right – as a kind of perpetual 'readiness'.*

Marynia Farnham and Ferdinand Lundberg (1947, p.142) in their influential tract, **Modern Woman: The Lost Sex,** remark that the 'traits necessary to the attainment of sexual pleasure' in women include

> *receptivity and passiveness, a willingness to accept dependence without fear or resentment, with a deep inwardness and readiness.*

Part of the pleasure psychoanalysts such as Helene Deutsch and Marie Bonaparte saw in female sexuality was 'surrender' to the male 'as testimony to the woman's own desirability' (quoted in

Poster for **How to Marry a Millionaire** (1953)

Ryan, 1975, p.283). We have already seen how much Monroe's image came to be understood in terms of her offering herself, making herself ready. In pin-ups, she often pushes herself forward from the shoulders; in the first shot of her in **Niagara,** she is lying on a bed in silk sheets which cling to her legs so that we can see they are open; in the ads for **How to Marry a Millionaire** it is she who beckons the viewer into the film. I've quoted above some of Monroe's statements that served to reinforce this notion that female sexual pleasure resided in part in being found pleasing.

Being found pleasing is part and parcel of the way that women, not men, are constructed as 'the beautiful sex' (see Una Stannard's article in Gornick and Moran, 1971). Though male sex appeal had long been a marketable commodity, it was still an oddly unspoken one. If a man was attractive to women, it was generally discussed in terms of qualities of personality that were only in part to do with how he treated women; whereas desirability in women had to do almost entirely with how a woman responded to men. Women showing an active sexual interest in men were generally labelled, popularly and psychoanalytically, as predatory or neurotic. Since women were the beautiful ones, there was perhaps a logic in representing female desire as being turned on not by men, but by their own bodies, as narcissism. From a psycho point of view, narcissism is both a normal aspect of female sexuality, and an aberration or kink. This is characteristic of how psycho constructs the female – what makes her normal is also what makes her peculiar and therefore, implicitly, a problem or enigma for men.

There is a considerable emphasis on narcissism in Monroe's image. She is often shown caressing herself. One of the most reproduced stills from any of her films is a shot that lasts only a few seconds in the film itself, **How To Marry A Millionaire** – but it catches the idea of her narcissism both in the pose, one hand thrown back behind her head in self-conscious abandon, and in the multiple mirrors, offering her an orgy of delight in her own reflection. Magazine stories were fond of playing up this angle too:

> *When she's alone, she often strikes art poses before a full-length mirror, admiring the beautifully distributed 118 lbs that millions of moviegoers admire (***Time** 11.8.52).

Readiness and narcissism both work aspects of male sexuality into female sexuality; but there was also a discourse on female sexuality that sought to establish it as something distinct on

Marilyn Monroe in **How to Marry a Millionaire** (1953)

its own: not the autonomous female sexuality to which the seventies women's movement has re-laid claim, but rather a sexuality implicitly dependent on men yet also quite different from male sexuality.

Clinically, in the psychotherapeutic theories and practices, this centred on the notion of the vaginal orgasm.* Kinsey's report on women had contributed to the growth of emphasis on female sexual satisfaction, but in one crucial regard his findings had little impact. His evidence pointed ineluctably to the clitoris as the organ of female sexual pleasure, but, as Regina Maxwell Morante points out, this evidence was systematically ignored in both the intensive media coverage of the report and in the psychotherapeutic world of counselling, etc: 'the persistent influence of Freudian theory on definitions of female nature and the absence of an organized feminist movement in the early 1950s softened the impact of this most radical of Kinsey's findings'. (Morante, 1977, p.574).

*One other dominant feature of the psycho discourse was masochism, which is not however an aspect of Monroe's image. Her desirability is bound up with vulnerability, which potentially overlaps with masochism, with pleasure in being hurt. However, if we are invited to take pleasure in her pain, her image nowhere signals that she might get pleasure from it too.

Psychoanalysis insisted that the clitoral orgasm was 'immature', or even, in Helene Deutsch's word, 'malicious'; and that only the vaginal orgasm was 'real' and 'truly satisfying'. But what was this orgasm like? Since women were supposed to achieve it yet somehow had great difficulty in doing so, it had to be described so that they would know what they were after. But how?

The experience of male orgasm can be identified with both the actions of the man in intercourse (his orgasm typically thrusts, beats, stabs, sears, grinds, and so on) and with the visible sign of it in the ejaculation of semen (which, in pornographic movies, is de rigeur as proof that you have 'really seen' an orgasm). But in the psycho discourse, with women there are neither actions nor visible product. As a result the orgasmic vocabulary becomes vague, formless, mysterious. At the level of psychotherapeutic theory, Marie Robinson's description of female orgasm as 'like going over Niagara falls in a barrel' is too extraordinary not to quote, but Anais Nin's description is more revealing. Nin has been hailed as a pioneer of female literary erotica, yet her description of 'the two kinds of orgasm', with the first, and vaginal, the preferable, is revealing:

> *[the first] in which the woman lay passive, acquiescent and serene. [It] came out of the darkness miraculously, dissolving and invading. In the other a driving force, an anxiety, a tension which made the woman grasp for it . . . [and] which brought an orgasm that did not bring calm satisfaction but depression* (quoted in Ryan, 1975, p.281).

Inevitably it is oceanic imagery that predominates in the popular literature. For instance, in **Return to Peyton Place** (1960):

> *Like a sea's retaining wall she lay and allowed herself to be buffeted, and felt the tidal pull that, at the end, seemed to draw her soul out of her body* (Metallious, 1960, p.182)

Characteristic of all these descriptions is that although notionally located in a specific place (the vagina) they evoke an experience that suffuses the whole body in a buffeting, dissolving, waterfalling ecstasy. Where the visible/visual analogue for the male experience derives from the penis, for the female it is everywhere. The visual analogue of the vaginal orgasm is the female body itself.

It is this that I want to argue in relation to Monroe. The presentation of her body, broadly in line with other traditions in Western culture but not in line with previous screen sex goddesses, is the analogue of the conception of female sexual experience that is expressed in the psycho discourse as the vaginal orgasm.

The most suggestive term through which to approach this is proposed by Mary Ellmann (1970, pp.74) in her book

Thinking About Women, namely formlessness:

> *The impression of women's formlessness underlies the familiar, and often most generous, acknowledgement of their superficial form . . . The flesh of women (as Sade would put it) is less resistent and less muscular than that of men. Pinched, it bruises more easily . . . Solid ground is masculine, the sea is feminine.*

Margaret Walters, in her study of **The Nude Male,** likewise points to a fundamental distinction throughout Western art between male and female form. She compares two Greek statues, of a man and a woman, both in a similar posture – yet the male nude suggests potency and an upwards reaching movement, the female passivity, a shrinking downwards. Even the physical texture of the bodies, exaggerated by the different materials used, contrasts: the male body is tautened, the anatomy outlined; the female body is slack, with a textured surface that blurs its outline. I emphasise the word 'outline' deliberately. In a discussion of the nature of form in post-Renaissance art, Peter Fuller (1980) argues that outlining things is the analogue of a rationalist view of the world that creates a separation between the world and the spectator – outline separates things off from one another, it is analytical, and at the same time, it also keeps the spectator at an 'objective' distance. (It is only with Cezanne and other early modernist painters, argues Fuller, that we begin to get a break-up of these kinds of distinction, in art at any rate.) This rational, distanced orientation to the world is surely also the public world of men.

My argument so far is in two stages. If traditionally a woman's body is held to be the analogue of the inner self, then, in a period where the inner self and sexuality are identified as one, the female body signifies ideas of female sexuality. And if the conception of female sexuality dwells on its dissolving, oceanic nature then the visual form it takes is, paradoxically, formlessness, slackness, blur. Yet it has not quite been the case that the sexuality female stars' bodies signified was the soft, blurred, 'vaginal' kind indicated by Ellmann, Walters and Fuller (in my use of them). It's not that women have not been represented with soft bodies, they have; but whenever the question of *their* sexuality is implied, the imagery becomes tougher, harder, more 'masculine', more 'phallic'. Supposedly male secondary characteristics are emphasised – wide shoulders (Greta Garbo, Joan Crawford), deep voices (Marlene Dietrich), slim hips and flat chests (the flapper, such as Clara Bow). Often their bodies are dressed in a style that 'hardens' them – Mae West's corsetting, Jane Russell's brassières, and so on. In terms of personality image, the earlier sex goddesses are either foreign (actually or supposedly) and therefore of a sexuality beyond all scrutiny (Theda Bara, Dietrich, Garbo, Anna May Wong), or else they are predatory, like a man (Bow, West, Jean Harlow). Obviously each star's image is more complex and contradictory than this, but I can think of none who quite incarnates an explicitly sexual image (sexual for the woman herself that is) that is also a soft and blurry one (though Betty Grable, Rita Hayworth and Margaret Sullavan clearly have something of that).

Even Monroe did not start out like that. The standard early pin-ups and the way she is framed and posed in the early films often do present an outlined form – albeit a crooked, rather than a straight, 'male' one. The reinforced fabric of the one-piece bathing costume and the side-on tits and arse positionings already discussed do not create a blurred, formless look. But already the **Golden Dreams** calendar is moving in that direction. Its tones are more soft and dissolving than the much more precise chiaroscuro of 'art photography', its pose more rounded and languorous than their sharp, contorted positionings. Three of the most famous sexy moments in her films show the progression to a more formless form as does her most characteristic facial pose – and we also have the unwitting testimony of a key witness to Monroe, Norman Mailer.

The three moments in question are the walk in **Niagara,** the subway gratings shot from **The Seven Year Itch,** and the bat and ball sequence in **The Misfits.** The walk in **Niagara** is a wiggle, invariably described by the critics as undulating, serpentine, squiggling, squirming, wriggling, a veritable thesaurus of terms connoting movement that cannot be determined (pinned down), that has no edges and boundaries. Similarly the subway scene, though Monroe keeps her body stiff, emphasises her billowing, undulating skirt. In the sequence in **The Misfits** she wears a loose spotted dress that holds her breasts in a softly moulding way, the antithesis of the hard-shaped, pointed breasts of brassière days; and the low camera angle catches the pendulous swing of her breasts as she bats the ball. In each case her body is rendered oceanic, unstable, vaginal.

Nowhere is this more true than in her most characteristic facial expression, repeated in every film, endlessly reproduced, an expression that even in a still photograph suggests movement, and was well enough described by **Time** as 'moist, half-closed eyes and moist, half-opened mouth'. The repetition of 'moist' hints at the obvious vaginal symbolism here, but that wetness isn't the

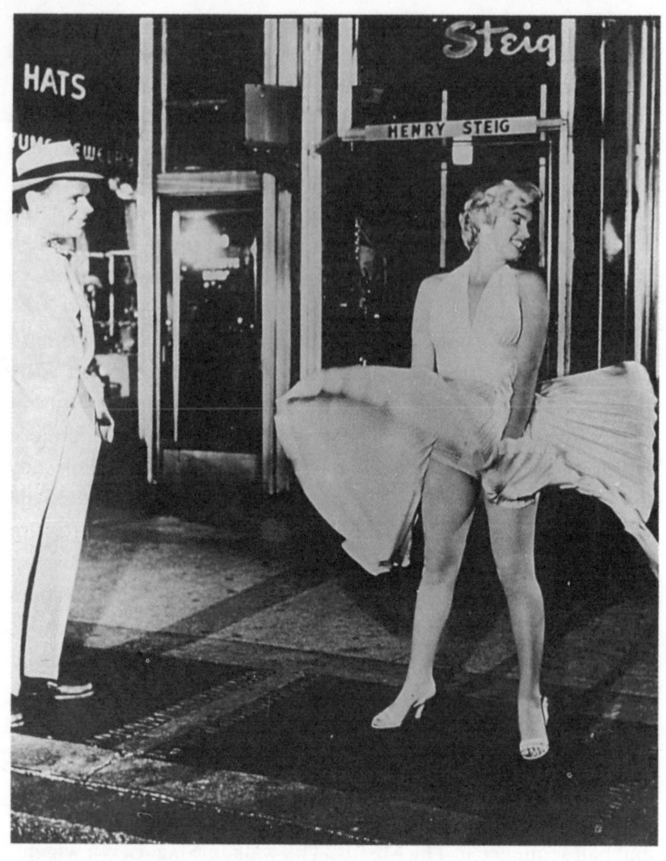

Tom Ewell and Marilyn Monroe in **The Seven Year Itch** (1955)

only thing that suggests it. Monroe's mouth, open for a kiss, is never actually still, but constantly quivering. Thus it forms not the neat round hole of Clara Bow's puckered lips or the hard butterfly set of Jean Harlow's mouth, but a shape that never actually takes on a definite shape, that remains formless.

My argument is not that quivering lips must be vaginal symbols, but that they are of a piece with other formless aspects of her image which together can be read as the visual analogue for a basic conception of female sexuality as itself formless. But don't take my word for it, take Norman Mailer's. The language of Mailer's book on Monroe constantly evokes her in 'vaginal' terms. Everything about her is soft. She is 'soft in her flesh', with 'a womb fairly salivating in seed' (Mailer, 1973, p.16), and her eyes are 'as soft . . . as a deer's' and her kisses are 'like velvet' (pp.45-6). In an extraordinary mixture of barely disguised sexual metaphors, he describes her in **Gentlemen Prefer Blondes** as both a flaccid (formless, soft) penis and a sweet, gaping vagina:

she seems close to a **detumescent** *body ready to roll right over the edge of the world and drop your body down* **a chute of pillows and honey** (p.102, my italics).

Finally she reaches her apotheosis, for Mailer, in **The Misfits,** and his description could not corroborate more clearly what I have been arguing about sexuality and formlessness:

She is not sensual here but **sensuous** . . . *she seems to possess* **no clear outline** *on screen. She is not so much a woman as a mood, a* **cloud** *of* **drifting senses** *in the form of Marilyn Monroe* (p.193 – all emphases save first added).

In terms of progress in images of women, the 'vaginal' representation of Monroe's body is ambivalent. Cast within the psycho discourse, the vaginal orgasm implies penetration and it's a sufficiently unimaginative discourse to equate penetration with insertion of the penis; in this regard as in regard to readiness and narcissism, the psycho discourse makes female sexual pleasure dependent on male sexuality, while denying the autonomous sexuality of the clitoris. In embodying that, Monroe is caught in it. Yet the development of a vaginal imagery also presages certain developments in feminist art in the seventies, where women artists began producing work that sought to reclaim knowledge and celebration of the vagina, that had for so long been despised as 'cunt', just there for men. This work has been criticised from within the women's movement as seeming to posit the vagina as the centre of women's being (itself a notion with a long history in male thinking about women). Yet at a certain point in time, such work was felt necessary as an attempt to claim a part of women's bodies for women's own consciousness, to wrest it from its definition as a function of male sexuality. Perhaps Monroe's 'vaginalism' did something to break up a system of sexual representation which could only comprehend sexuality by using male sexuality as the model. But this example already points to how difficult it is to discuss Monroe's image in the light of the contemporary women's movement.

Sex Pol

I have been trying to read Monroe in the discourses current at the time she was a working star. The examination of the ways in which Monroe's image has been taken up and articulated in relation to changing patterns of ideas in the sixties, seventies and into the eighties is a whole different enterprise. Here I want to use feminist perspectives to get at what is, after all, there to be read in the image, that not yet quite articulated sense that all is not right with the position of women in contemporary capitalism, Betty Friedan's 'problem that has no name', Brandon French's 'verge of revolt'.

The contemporary women's movement has seen Monroe as, at worst, the ultimate example of woman as victim as sex object, and, at best, as in rebellion against her objectification. Susan Brownmiller (1975), quoted above, is one of the clearest expressions of the former position; while Gloria Steinem's (1973) article, 'Marilyn – the Woman Who Died Too Young' in **Ms** magazine, was one of the first statements of the latter position. Yet Steinem makes it clear that this is retrospective – 'We are taking her seriously *at last*' (my italics) – and Molly Haskell (1974, p.254) makes the point directly:

> *Women . . . have become contrite over their previous hostility to Monroe, canonizing her as a martyr to male chauvinism, which in most ways she was. But at the time, women couldn't identify with her and didn't support her.*

I haven't done an audience survey, but I have presented this material to many groups with women who remembered Monroe as a star when she was alive, and by far the majority confirm what Haskell says. This is not to say that Monroe was never an identification or sympathy figure for women; but it is to say, at the least, that the terms in which she may have been, and in which her image articulates the question of being a sex object, were *not* the terms of objectification, desire, and so on, that became familiar during the seventies. The most we can say is that her image did at certain points crack open these discourses, but there was nothing to draw on to construct the kind of elaborated discourse on woman as sexual spectacle that feminism has since given us.

Two widely circulated quotations may illustrate this. At first glance they seem to come very close to the feminist woman-as-sex-object critique. In her first press interview after the break with Hollywood, she is quoted as saying:

> *I formed my own corporation so I can play the better kind of roles I want to play. I didn't like a lot of my pictures. I'm tired of*

sex roles. I don't want to play sex roles any more (Zolotow, 1961, p.191);

and in her last interview, for **Life,** she said:

That's the trouble, a sex symbol becomes a thing – I just hate to be a thing.

It's possible to pick quotations out like this and make it seem that Monroe was articulating a prescient seventies feminist position, the alienation (thing-ness) of being treated so relentlessly in terms of sex.

But the context confuses this reading – or rather, suggests how necessarily confused any such articulation is. In the case of the first, Monroe's appearance at the press conference seemed to undercut what she was saying. Here is Maurice Zolotow's (p.191) description, characteristically prurient, but also indicating how much Monroe's declarations were compromised by her apparent self-presentation:

the 'New Marilyn', when she arrived, didn't look as if she were a candidate for a convent, or even for the role of one of Chekhov's sisters. She swished into the salon in a tight white satin gown which revealed forty per cent of her bosom, she was made up to the nines – hair in a new shade of 'subdued' platinum and lips carmined to extravagance.

Such self-objectification accompanying a refusal to be objectified was generally read in favour of the former, with the latter ridiculed in line with her dumb blonde image. Another way of seeing it is that presenting herself as a sex object was at least guaranteed to bring recognition and approval – the gap between how she looked and what she said enacts the tension involved in asserting a sexual critique when no such critique yet exists.

Similarly, it's important to read the rest of the **Life** quote:

But if I'm going to be a symbol of something, I'd rather have it sex.

Time and again, Monroe seems to buy into the 'progressive' view of sex, a refusal of its dirtiness – but that means buying into the traps of the sexual discourses discussed above: the playboy discourse, with women as the vehicle for male sexual freedom, and the psycho discourse, with its evocation of the ineffable unknowability of sexuality for women. The choice of roles from **Bus Stop** on indicates the conundrums the image is caught up in. All of them are still characters defined in terms of their sexual meaning to men. Only **The Misfits** begins to hint at a for-itself female sexuality, and then casts it utterly within the discourse of female sexuality as formlessness. The men in the film look on, unable to comprehend her sensuality; grasping a tree she looks out at them/us with a hollow expression of beatitude, straining to express what is already defined as inexpressible.

Yet some of her later films do contain hints of the struggle, traps and conundrums of living the fifties discourses of sexuality. **Bus Stop** is probably the most extended example. It is possible, without straining too much against the drift of the film, to read Cherie/Monroe not just as the object of male desire but as someone who has to live being a sex object.

Cherie/Monroe's longest dialogue comes in scenes with other women characters, and has thereby the quality of unburdening herself rather than putting on an act or standing up to men. With the waitress Vera (Eileen Heckart), she expresses her ambition to be a singer and Hollywood actress, referring to the lack of respect she has had up to now. With the young woman on the bus, Elma (Hope Lange), she speaks of the ideal man she is

Marilyn Monroe in **The Misfits** (1961)

looking for, a notion of a man who combines traditionally masculine and feminine qualities:

> *I want a guy I can look up to and admire – but I don't want him to browbeat me.*
> *I want a guy who's sweet to me, but I don't want him to baby me.*
> *I want a guy who has some real regard for me, aside from all that loving stuff.*

None of this is desperately radical or progressive; the ambition is mainstream individualism ('I'm trying to be somebody'), the ideal man is a sentimental fantasy. But both are located in the consciousness of a dumb showgirl type, and given a legitimate voice by the seriousness of the performance, of the way in which they are filmed, and, especially, by virtue of the fact that they are spoken to another woman. The ambition and fantasy are not in the slightest ridiculed, and they have the effect of throwing into relief the showgirl role that Cherie/Monroe is playing.

This throwing into relief also works in aspects of performance, notably surrounding the *Old Black Magic* number. The choice of this is appropriate anyway – Beau is looking for an 'angel', and when he enters the saloon and sees Cherie for the first time he lets out, 'Look at her gleaming there, so pale and white'. She is his luminous angel – but the song is *That Old* **Black** *Magic*, suggesting the opposite of the angelic. As the number proceeds we see how she is producing her image. She kicks light switches to change the colour of light in relation to the words of the song; and in close-ups we see her putting on an act, adopting an excessive classical posture for 'I should stay away but what can I do?' and exaggerated hand-to-mouth kissing gestures for 'kiss, kiss, kiss'. Before the first of these, she winks at Beau and smiles in a giggly way, implying a collusion with him in putting on this act.

Yet we must set all of this dialogue and performance against a lot else that is happening in the film, and more insistently as it goes on. Cherie/Monroe is the butt of three sight gags that we are clearly meant to find funny, though they humiliate her and endorse male supremacy. The first occurs at the rodeo, when a journalist and photographer from **Life** come to take her picture (because Beau, from the arena, has just announced their marriage). She bends over to pick up her lipstick from her bag and the photographer snaps a picture of her backside, and goes. 'But I didn't have my lipstick on yet,' she wails. At one level, this could be read as Cherie/Monroe wishing to be a glamorous object, but not quite in the dehumanising terms of the photographer; but equally, the gag (especially the line) only works as a gag if we collude with the photographer, with the assumption that of course we'd rather see her backside than her face and only a dumb broad wouldn't realise that. Later, in the saloon, when Cherie tries to get away from Beau he grabs her costume and its tail comes off in his hands – 'Give me back my tail,' she screams. Again, this could be mocking the absurd accoutrements of the show girl, but the line, because she refers to it as 'my tail', mocks her. The third gag (already discussed above) occurs at the bus station, where Beau lassoes her as she is trying to get on a bus to escape him. We are invited to laugh at least despite this gross humiliation, and probably at it.

Elements of dialogue and performance seem to open the show-girl image up for scrutiny while the gags (and the ultimate working out of the story) invite us to laugh at her (and endorse her final capture).

This is to suggest that the film is now one thing (crypto-feminist), now another (male chauvinist) – or a confusion of both. The way in which Cherie/Monroe is to-be-looked-at, as the film constructs it, illustrates this well, and perhaps gets most exactly the sense in which – and the limit to which – the film/Monroe is opening up the discourse that the resurgent women's movement would finally articulate.

The first shot of Monroe, as so often in her films, is a point-of-view shot. But it is not Beau, the hero's, point of view but Virgil, his mentor's. Indeed, as the latter looks out of the hotel bedroom window, the film cuts to her (Virgil's point-of-view) and then back to Beau shaving, underlining the fact that he does *not* see her (and will not until the saloon scene). This means that the shots of her that follow can articulate more than Monroe as an object of desire, because they have not been set up as contained by desire in the first place.

Virgil's point-of-view shot is a long shot, and it is followed by a mid-shot (at the same looking-down-at-her angle), then a return to the long shot. In this way, we get to see her better than Virgil does. In one sense, this satisfies our voyeurism – both sexual (we have a better view than Virgil) and in terms of narrative, character and star (we want to know who she is, what Monroe looks like in this part, and so on).

In the second long shot, a group of men enter the room, crowding around her as she tries to fend them off. They are broken up by the saloon owner who clears them out and then yells at her to get back to work. The image very clearly sets out the dimensions of male power (of the male audience/clients, of the male employer) within which Cherie/Monroe is caught. She is also caught in our/the camera/Virgil's gaze, but what we see articulates something of what it is like to be gazed at. We gaze at an image that hints at the politics of gazing.

The scene that follows takes place inside the dressing room, no longer seen by Virgil. It allows us to see Cherie/Monroe close to, and to observe what we could not in the opening shots, which preserve something of the magic and beauty of the half-dressed woman glimpsed from a hotel window. Her hair looks as if it has been peroxided (it would not convince us that she was a natural blonde); her face looks deathly white; her stockings have holes in them. This deglamourising continues in the ungainly way she gets into her green sateen leotard. Scenes in showgirls' dressing rooms are usually voyeuristic – **The Prince and the Showgirl** plays with our expectations of 'seeing something we shouldn't' by teasing camera movements and set-ups, delaying our actually seeing Elsie/Monroe till she has just finished dressing. The treatment in **Bus Stop** is, by contrast, head-on. There is nothing we don't see (within the conventions of the period), but there is no sense of our being there just to see something, no tease to lead us on as in **The Prince and the Showgirl.** We *are* there to see it, it is true, and the fact that we are seeing Marilyn Monroe getting into a leotard must lend itself to the pleasure of voyeurism, but the ungainliness, the matter-of-fact conversation with a woman who isn't a showgirl, the tacky setting, none encourages this way of looking. Moreover, the movement shows us the showgirl setting up her act and the conversation stresses wanting to get beyond doing this kind of thing, so that within an image that is traditionally set up for the pleasure of gazing, we are again getting some explorations of what it means to be someone who lives by being gazed at.

The further we get from Monroe and the fifties, the more it seems that her image is so malleable that it can mean almost anything. Both Molly Haskell (1974) and Brandon French (1978), in their discussions of **Some Like It Hot,** speak of Monroe being androgynous, part of the continuum of gender identity confusions of the film that includes men in drag, all-girl jazz bands, men marrying men and so on. Monroe as androgyne is something else again from Monroe as the most womanly of women. When writing the section on Monroe and vaginal imagery, my mind kept slipping between seeing her as embodying notions of sexual (vaginal) fulfilment and as grasping after a sexual fulfilment that constantly eludes her, her mouth always ready but with no signs of satisfaction. I suspect, however, that our latter reading is informed by the assumption (itself complicated) that the image is reaching out to embody an experience that does not in fact exist. In the context of a belief in the vaginal orgasm, there is no reason why her movements and expressions should not signify and incarnate that sense of female sexual fulfilment.

What meanings Monroe can and does carry today would have to be approached through the discourses (such as the interest in androgyny, or the debates about the clitoral versus the vaginal orgasm) that have been constructed in the twenty odd years since her death. Why she should be able to articulate them is in turn an interesting question. Perhaps it is because she can be a talisman of what we are rejecting, of the price people had to pay for living in the regime of sexual discourses of the fifties. She flatters our sense of being so advanced. But perhaps too we are not so far from the fifties as we might like to think – notions of natural sexuality, of repression, of the ineffability of female sexuality, of sexuality as the key to human happiness and truth, these are not notions we have left behind. As long as sexuality goes on being privileged in quite the way it is, Monroe will be an affirmation of that principle even while also being witness to the price we pay for it.

Paul Robeson: Crossing Over

One ever feels his twoness – an American, a Negro; two souls, two thoughts, two unreconciled strivings; two warring ideals in one dark body, whose dogged strength alone keeps it from being torn asunder.

W.E.B. DuBois, **The Souls of Black Folk**

In the jargon of the contemporary pop music business, a cross-over star is one who appeals to more than one musical subculture; one who, though rooted in a particular tradition of music with a particular audience, somehow manages to appeal, and sell, beyond the confines of that audience. Dolly Parton, Gladys Knight, Paul McCartney are recognisably country, soul and rock performers respectively, but they have a following among people who are not especially into those kinds of music. While having this wider appeal, they are still rooted in the particular musical subculture that defines them – in crossing over, they don't lose their original following. Or not too much of it.

The term cross-over, in this sense, did not exist when Paul Robeson was a major star, but, at least between 1924 and 1945, he was very definitely an example of it. His image insisted on his blackness – musically, in his primary association with Negro folk music, especially spirituals; in the theatre and films, in the recurrence of Africa as a motif; and in general in the way his image is so bound up with notions of racial character, the nature of black folks, the Negro essence, and so on. Yet he was a star equally popular with black and white audiences. There were other black singing stars as, if not more, popular than he in the twenties and thirties – Louis Armstrong, Bessie Smith, Ethel Waters, Cab Calloway, Billie Holiday – but none of these quite established the emblematic or charismatic position, for blacks and whites, and in more than one medium, that Robeson did. How did he manage this?

Some would argue that his achievements were so unarguably outstanding that he had to be recognised. Certainly there is no gainsaying those achievements – a brilliant academic record at Rutgers University (1915–19), where he was the only black student at the time and only the third ever to have been admitted, and at Columbia graduate law school (1920–1) where he was again the only black student; a great football player, the first black player ever selected to play for the national team (the All-Americans) from the university teams, all the more remarkable, according to Murray Kempton, for being from Rutgers, a less prestigious university (Kempton, 1955, p.238); certainly the best known and most successful male singer of Negro spirituals, in concert and on record, and always highly acclaimed critically; the performer of what has been called the definitive Othello of his generation ('The best remembered Othello of recent decades', wrote Marvin Rosenberg in 1961 (p.151)) and in the longest running Shakespeare production in Broadway history (1943-4); one of those performers who has made one of the standards of the show business repertoire, *Old Man River,* wholly identified with him, so much so that he was himself often referred to as Old Man River; singer of the hugely successful patriotic *Ballad for Americans* (1939) that the Republican party adopted for their National Convention in 1940; rapturously received in the theatre, particularly in **The Emperor Jones** (New York, 1924; London, 1926; Berlin, 1929), **All God's Chillun Got Wings** (New York, 1924; London, 1933), **Show Boat** (London, 1928; New York, 1932) as well as **Othello** (London, 1930; New York and US tour, 1943-5; Stratford-on-Avon, 1959); and if nothing in his film career quite matches up to all that, still he was big enough to be billed outside a London cinema for the première of **Song of Freedom** (1936) as 'GREATEST SINGING STAR OF THE AGE' and worked with the most important black film-maker of the time, Oscar Micheaux (in **Body and Soul,** 1924), with the avant-garde group surrounding the magazine **Close-Up (Borderline,** 1930), in a lavish Hollywood musical (**Show Boat,** 1935) and in a series of popular and, shall we say, for the most part decent British films of the thirties, as well as narrating a number of documentaries, including **Native Land** (1941) by Frontier Films and, outside our period, **Song of the Rivers** (1954) by Joris Ivens.

Yet such a list of achievements does not really explain his stardom. Whilst I don't want to diminish his talent or effort, this list is for the most part a statement of the fact that he was highly acclaimed, very popular, in other words, a star. But why were *these* achievements of so much interest to so many people?

How were his remarkable qualities also star qualities? How and why were these the qualities of the first major black star?

We need to get the question right. How did the period permit black stardom? What were the qualities this black person could be taken to embody, that could catch on in a society where there had never been a black star of this magnitude? What was the fit between the parameters of what black images the society could tolerate and the particular qualities Robeson could be taken to embody? Where was the give in the ideological system?

Yet another way of putting it might be – what was the price that had to be paid for a black person to become such a star? Harold Cruse, the subtlest critic of both Robeson and the Harlem Renaissance with which he was associated, argues that from the perspective of black politics and black consciousness the price was too high. He finds Robeson too integrationist, too concerned with adapting himself to white cultural norms, too far removed from the real cultural concerns of black people, and too little aware that cultural development is not a thing of the spirit alone but is rooted in material conditions, the necessities of funding and support usually absent in black communities (Cruse, 1969, 1978). Likewise, Jim Pines (1975, p.32) suggests that Robeson's work is a 'largely individualist and generally mystifying protest . . . [that] seems to substantiate the ineffectualness of individualist forms of protest against cultural exploitation by the media'; and Donald Bogle (1974, p.98) even argues that Robeson was not only individualist in Pines' sense of an isolated individual but in the sense of self-seeking – 'No matter how much producers tried to make Robeson a symbol of black humanity, he always came across as a man more interested in himself than anyone else'.

The figure of Robeson still sets off argument about images of black peoples, and there is still a striking disparity in the different ways he is perceived. I am confining myself to the period in which Robeson was a cross-over and major star, roughly 1924–45, when the politics are less explicit and explosive than afterwards. What particularly interests me is the way that Robeson's image takes on different meanings in this period when read through different contemporary black and white perspectives on the world. There are different, white and black, ways of looking at or making sense of Robeson, but it would be a mistake to think that the white view is the one that stresses achievement, the black the one that stresses selling out. To set against Cruse, Pines, Bogle and the other black writers, there is **Paul Robeson: The Great Forerunner,** a book of articles and tributes, produced by the radical black magazine **Freedomways** (**Freedomways** editors, 1978), as well as countless celebrations of Robeson by black people throughout his career. Likewise, one can find – less easily, it is true – white writers quick enough to point to his artistic limitations or, like Murray Kempton (1955, p.259), to deny his significance as an embodiment of black people: 'There was absolutely nothing between him and the people for whom he affected to speak'.

Equally, in speaking of different, white and black perspectives, I don't imply that black people saw him one way and whites the other. What I want to show is that there are discourses developed by whites in white culture and by blacks in black culture which made a different sense of the same phenomenon, Paul Robeson. There is a consistency in the statements, images and texts, produced on the one hand by blacks and on the other by whites, that makes it reasonable to refer to black and white discourses, even while accepting that there may have been blacks who have thought and felt largely through whites discourses and vice versa.

The difficulty of this argument is not so much theoretical as discerning the difference between black and white discourses in relation to Robeson. The difference is not obvious in the texts, and that is part of the explanation of how Robeson's cross-over star position was possible. For much of the time it could seem that the black and white discourses of the period were saying the same thing, because they were using the same words and looking at the same things. Robeson was taken to embody a set of specifically black qualities – naturalness, primitiveness, simplicity and others – that were equally valued and similarly evoked, but for different reasons, by whites and blacks. It is because he could appeal on these different fronts that he could achieve star status.

All the same, Robeson was working, particularly in theatre and films, in forms that had been developed, used and understood in predominantly white ways. In appearing in white plays and films, Robeson already brought with him the complex struggle of white and black meanings that his image condensed – but what happened to those meanings? Just by being in the plays and films, some of those black meanings are registered – but they are part of a broadly white handling of him, and this is significant not so much at the level of script and dialogue, as at the level of various affective devices that work to contain and defuse those black meanings, to offer the viewer the pathos of a beautiful, passive racial emblem.

The strategies of the white media worked to contain Robeson, but only *worked* to. How he was handled by the media is conceptually distinct from how audiences perceived him. By so doing the media might reinforce white discourses that intended to contain what was dangerous about black images in general, but by registering the black meanings in his image they also made these widely available, for use by black people. There is plenty of evidence of the impact that Robeson had on black audiences, but James Baldwin's notion of the way that Robeson (and other black stars) could work against the grain of his films suggests, at least, how a black viewer could see it that way, and with the force of 'reality' and 'truth':

> *It is scarcely possible to think of a black American actor who has not been misused: not one that has ever been seriously challenged to deliver the best that is in him . . . What the black actor has managed to give are moments, created, miraculously, beyond the confines of the script; hints of reality, smuggled like contraband into a maudlin tale, and with enough force, if unleashed, to shatter the tale to fragments . . . There is truth to be found in . . . Robeson in everything I saw him do* (Baldwin, 1978, pp.103–5).

It is the range of potential reading, black and white, in the context of an overall white media handling of Robeson that is the subject of this chapter.

Essential Black

Paul Robeson was widely regarded as the epitome of what black people are like. People felt that they could see this in the way he stood in concert, 'tightening his broad shoulders to bear a load of agonised entreaty, casting the outlines of his head into a sort of racial stereotype' (Sergeant, 1926, p.196), or hear it in his voice – 'it was Mr Robeson's gift to make [the spirituals] tell in every line . . . they voiced the sorrows and hopes of a people' (**New York Times** of his first concert, 19 April 1925). Ollie Harrington's (1978, p.102) memories of a black ghetto upbringing, in the Bronx and then in Harlem, equally testify to how deeply racially significant Robeson was felt to be, an example of the fact that black people were capable of achieving what whites could and also someone whose singing 'gripped some inner fibres in us that had been dozing'.

A similar feeling is evoked by Eslanda Goode Robeson in her 1930 (pp.67–8) biography:

Paul Robeson was a hero: he fulfilled the ideal of nearly every class of Negro. Those who admired intellect pointed to his Phi Beta Kappa key; those who admired physical prowess talked about his remarkable record. His simplicity and charm were captivating; everyone was glad that he was so typically Negroid in appearance, color and features . . . He soon became Harlem's special favourite, and is so; everyone knew and admired him . . . When Paul Robeson walks down Seventh Avenue . . . it takes him hours to negotiate the ten blocks from One Hundred and Forty-Third Street to One Hundred and Thirty-Third Street; at every step of the way he is stopped by some acquaintance or friend who wants a few words with him.

This passage is a major moment in the authentication of the Paul Robeson image. It authenticates his image as the Negro man *par excellence* by showing that it is based in grass roots approval in the heartland of urban black America, Harlem; and that is itself authenticated by being recounted by his wife. (At the level of popular hagiography, no niceties about spouses as unreliable witnesses intrude.) Eslanda's biography was a basis for much of the subsequent writing about Paul – this 143th to 133rd story recurs endlessly, for instance.

The early recognition of Robeson as the incarnation of blackness was developed in his subsequent career. He played or was associated with many of the male heroes of black culture. He sang *John Henry* in concerts and on records, and appeared in 1940, albeit briefly due to illness, in Roark Bradford's play about that more or less mythical worker hero of the 1870s. In 1941 he recorded, with Count Basie, *King Joe,* a tribute to Joe Louis, the most celebrated and successful of black boxing champions, and himself very much seen as the pride of his race (see Levine, 1977, pp.429ff). He had, moreover, already played a boxer, a kind of generic black hero figure, in **Black Boy** in 1926. He played the lead in C.L.R. James's play **Toussaint L'Ouverture** in London in 1936, and was to have played in Sergei Eisenstein's film based on the revolutionary leader. There are echoes of Shine, the legendary black stoker on the *Titanic* who saw trouble ahead and swam an ocean to safety (Levine, ibid., p.428), in Robeson's role as Yank in **The Hairy Ape** (1931) and the scene in the boiler room in the film of **The Emperor Jones** (1933). There is a portrait of Booker T. Washington behind the happy ending grouping of Robeson (as Sylvester) with wife and mother-in-law in **Body and Soul** (1924), and he was associated with the other major black intellectual

1936—The Classic Universal Version

Advertisement for **Show Boat** (1936)

leader of the times, W.E.B. DuBois, in public appearances at numerous meetings and rallies.

If he played or was associated with the heroes of black culture, he also played the stereotypes of the white imagination – Lazybones (the role in **Show Boat**), smiling Sambo (the ads for **Show Boat,** Bosambo in **Sanders of the River,** 1934), variations on the plantation/sharecropping Jim Crow (**Shuffle Along,** 1920; **Voodoo,** 1922; **Tales of Manhattan** 1942), black nobility (**Song of Freedom,** 1936; **Big Fella,** 1937), creature of the ghetto (**Rosanne,** 1924; **Body and Soul,** 1924; **All God's Chillun Got Wings,** 1924; Crown in **Porgy and Bess,** 1927; **The Emperor Jones**) and brute (Crown, Brutus Jones, **Voodoo, Black Boy, The Hairy Ape, Basalik,** 1935; even **Othello** and **Stevedore,** 1935) – to name but a few.

Small wonder then that Robeson should be so identified as the representative of blackness. Apart from the brute stereotype, these images of blackness, of Paul Robeson's blackness, were, whatever we may think about them now, affirmatively valued in the discourses of the period. Robeson represents the idea of blackness as a positive quality, often explicitly set over against whiteness and its inadequacies.

Typical of the general view of blackness that Robeson

Paul Robeson (l.) in **Shuffle Along** (1920)

represented are statements like the following, from a white enthusiast for black culture:

> *It is . . . the feeling for life which is the secret of the art of the Negro people, as surely as it is the lack of it, the slow atrophy of the capacity to live emotionally, which will be the ultimate decadence of the white civilised people* (Mannin, 1930, p.157).

Such sentiments are part of a general revulsion with contemporary industrial society, of which Mabel Dodge's (1936, p.453) words are typical:

> *America is all machinery and money making and factories. It is ugly, ugly, ugly.*

(Dodge was the patron of the Greenwich Village circle that had close links with the artists of the Harlem Renaissance.) Much of the work Robeson was associated with, and especially Eugene O'Neill's dramas, plays on this opposition of basic black and white racial/cultural differences. The stage directions for **All God's Chillun Got Wings** spell it out quite precisely:

A corner in lower New York . . . In the street leading left, the faces are all white; in the street leading right, all black. It is hot spring . . . People pass, black and white, the Negroes frankly participants in the spirit of spring, the whites laughing constrainedly, awkward in natural emotion. One only hears their laughter. It expresses the difference in race (Act One, Scene 1).

John Henry Raleigh (1965, p.110) notes of the black/white symbolism of **The Hairy Ape,*** and particularly the central scene between 'the terrified slender white-skinned, white-clad Mildred and the black, half-clad, muscular brute, Yank', that it suggests a 'familiar theme in the American racial situation . . . namely, that black stands for animal vitality, while white signifies frayed nerves'.

*Robeson was in a revival of this in 1931. The part was not written as a black one, but the symbolism, including the use of the ape motif, made it seem very suitable for casting Robeson.

These are all positive evaluations of blackness by whites, but one can find similar statements about blackness by black writers. The words are not exactly the same, and behind minor differences in wording lie very significant differences in understanding, but they seemed to be saying the same thing. The locus classicus of the black view of blackness is **The New Negro,** a collection of essays edited by Alain Locke and published in 1925, which is often seen as the manifesto of the Harlem Renaissance. Whatever the topic, the same kind of evaluation of black (or, as they prefer it, Negro) art and culture recurs. Compare Ethel Mannin's 'Negro gift for life' and 'atrophy of white civilisation' with Albert C. Barnes on Negro art:

[Negro art] is a sound art because it comes from a primitive nature upon which a white man's education has never been harnessed . . . the most important element . . . is the psychological complexion of the negro . . . The outstanding characteristics are his tremendous emotional endowment, his luxuriant and free imagination . . . The white man in the mass cannot compete with the Negro in spiritual endowment. Many centuries of civilisation have attenuated his original gifts and have made his mind dominate his spirit (Locke, 1968, pp.19–20).

Or compare Mabel Dodge's disgust at industrial, money-mad ugliness with J.A. Rogers on jazz:

The true spirit of jazz is a joyous revolt from convention, custom, authority, boredom, even sorrow – from everything that would confine the soul of man and hinder its riding free on the air . . . it has been such a balm for modern ennui, and has become a safety valve for modern machine-ridden and convention-bound society. It is the revolt of the emotions against repression (ibid., p.217).

And so on.

Paul Robeson himself was one of the clearest exponents of this view. In an article in **The Spectator** of 15 June 1934, he spelt out this black/white, emotion/intellect, nature/civilisation opposition very directly:

> *The white man has made a fetish of intellect and worships the God of thought; the Negro feels rather than thinks, experiences emotions directly rather than interprets them by roundabout and devious abstractions, and apprehends the outside world by means of intuitive perceptions instead of through a carefully built up system of logical analysis* (Robeson, 1978, p.65).

One difference between some of the black discourses on blackness (and Robeson) and some of the white is the kind of relationship assumed between the spontaneous/natural/simple/emotional quality of blackness and the civilised/rational/technological/arid quality of whiteness. Many of the contributors to **The New Negro** see the relationship very much in terms of black emotion as a resource to be transformed into a fully mature culture and to revitalise aesthetic expression. They often stress its limitations as it stands, but see it rather as a shot in the arm for pallid white art. Often too they have a vision of a synthesis of black feeling and white intellect, black sensuousness and white technology, and this emerges very clearly in their treatment of Africa. Two of Robeson's films, **Song of Freedom** (1936) and **Jericho** (1937) are explicit statements of this theme, both concerned with a Western black man who returns to (**Song of Freedom**) or lands up in (**Jericho**) Africa, and sees his mission as the bringing of the benefits of Western medicine, technology and education to the vibrant emotional life of the country. The films thus explore (more complexly than I've just indicated) one of what Alain Locke (1968, pp.14–15) calls the

> *constructive channels opening out into which the balked social feelings of the American Negro can flow freely . . . [namely] acting as the advance-guard of the African*

peoples in their contact with Twentieth Century civilisation . . . Garveyism may be a transient, if spectacular, phenomenon, but the possible role of the American Negro in the future development of Africa is one of the most constructive and universally helpful missions that any modern people can lay claim to.

The white positive valuation of blackness does not have these tendencies towards racial synthesis, but on the contrary often seems to want to ensure that blacks keep their blackness unsullied. One can see this in much of the white critical response to Robeson's Othello.

Robeson played Othello three times – in London in 1930, in New York and on tour from 1943–5, and at Stratford-on-Avon in 1959. It was well received, though English critics were protectively quick to say of the 1930 production that Robeson had not really mastered the verse. Still, James Agate, for instance, praised, without stating the racial connection, those qualities that reproduced an inflection of the notion of blackness we are discussing – by the end of the play, 'Othello ceased to be human and became a gibbering primeval man' (**Sunday Times,** 25.5.30; quoted in Schlosser, 1970, p.133). By the time of the New York production Robeson had 'mastered the verse'; he was also playing the role in terms of Othello as a man of dignity whose racial honour is betrayed (rather than purely in terms of sexual jealousy). The dignity and verse-speaking were noted, but the critics were now regretting that, for instance, his 'savagery is not believable, the core of violence is lacking' (Rosamund Gilder, **Theatre Arts,** 27 December 1943, p.702; quoted in Rosenberg, 1961, p.153). Similar comment greeted the Stratford performance, Alan Brien explicitly linking Robeson's Othello to the safe, gentle image of the black man and regretting the absence of the primitivistic element:

Mr Robeson . . . might be the son of Uncle Tom being taught a cruel lesson by Simon Legree . . . I pitied him, but in my pity I never felt any of the wild, guilty, apocalyptic exultation at the vision of Chaos come again (**The Spectator,** 10.4.59).

As Marvin Rosenberg (1961, p.202) shows in his study of the history of **Othello** productions, despite the many disagreements about the interpretation of the title role, the idea of its involving 'a passion so demented, so entire that once roused it seizes and dominates the man' has been held as essential since the play was first produced. This can be understood in purely individual terms – Othello is someone like that – or universal ones – everyone is like that – but it can also be understood racially. The play allows this rather forcefully – Iago is able to draw upon a fund of racist ideas about blacks, and black sexuality, to further his own resentment against Othello:

> (to Brabanto, Desdemona's father)
> *Even now, now, very now, an old black ram*
> *Is tupping your white ewe. Arise, arise . . .*
> *Or else the devil will make a grandsire of you.*
> (Act One, Scene 1)

and he articulates the claims of (white?) reason against (black?) sensuality:

> *If the balance of our lives had not one scale of reason to poise another of sensuality, the blood and baseness of our natures would conduct us to most preposterous conclusions. But we have reason to cool our raging motions, our carnal stings, our unbitted lusts.*
> (Act One,Scene 3)

We have no need to take the racial equation. Certainly the character of Othello might be played as a contradiction of Iago's view, the play as a whole not endorsing this concept of blackness. This presumably was part of what Robeson was trying to do. However, as Rosenberg notes (though again without pressing the evident connection with racial notions), it has been common in the twentieth century to view Othello in terms very close to the definition of blackness we are discussing – the idea of Othello 'as a *primitive* or barbarian veneered by civilisation, hence easily plunged into *savage* passion, has been a fairly popular critical interpretation' (Rosenberg, 1961, pp.191–2, my italics). Faced with a black man, Paul Robeson, playing Othello, white critics easily expected this primitivist interpretation and when they did not get it were disappointed (especially as they felt they had done so in 1930).

Robeson's Othello did make an issue out of the fact that Othello was black, or at any rate did so at the level of widely reported intention. But he wanted to emphasise the social position of Othello, a black man in a white society, someone living outside their culture. The critical reaction, however, suggests a desire to see the racial dimension in terms of essential racial differences, the blackness of emotionality, unreason and sensuality.

Black and white discourses on blackness seem to be valuing the same things – spontaneity, emotion, naturalness – yet giving them a different implication. Black discourses see them as contributions to the development of society, white as enviable qualities that only blacks have. This same difference runs through the two major traditions of representing blacks that Robeson fits into – blackness as folk, blackness as atavism.

Black as Folk

We black men seem the sole oasis of simple faith and reverence in a dusty desert of dollars and smartness.
W.E.B. DuBois, **The Souls of Black Folk**

In his biography of Jerome Kern (the composer of **Show Boat**), Gerald Bordman (1980, p.400) recounts a visit made by Kern to a Phil Silvers cabaret in 1944, in which there was a skit on Kern teaching Paul Robeson *Old Man River:*

> *Much of the fun derived from contrasting the grammar and rhetoric of the unlettered black man who is supposed to sing the song in the show with the college-educated Robeson's meticulous English. Robeson, for example, demands to know what 'taters' are and when he is told then attempts to sing the line in his impeccable Rutgers grammar. But even Robeson must admit that 'He doesn't plant potatoes . . .' fails to work.*

Part of the gag is anti-Robeson, his overeducatedness and effrontery in changing the words of the song (discussed below); but it does have a more progressive implication too.

No one, presumably, has ever thought *Old Man River* was a genuine folk song, but it has often been credited with having a folk feel and certainly as sung by Robeson it was widely felt to express a Negro essence. Virginia Hamilton, in her 1974 biography for children, writes of the London **Show Boat** audience listening to *Old Man River* and gazing 'openmouthed at this man who had somehow clasped the history of his people to his soul' (p.51). Yet what the Phil Silvers skit emphasises is the artificiality of this. Robeson as Joe was not a case of natural emanation but something that had to be worked up, learnt, produced as a particular image (and then, later, resisted as an image). The very idea of folk culture or consciousness is a construct, though one that gets its force and appeal from appearing not to be, from notions of naturalness and spontaneity.

In a discussion of black folk literature in **The New Negro** (Locke (ed.), 1968, p.242), Arthur Huff Fausset conveniently summarises the components of this concept of folkness. Folk is characterised by

1. Lack of the self-conscious element found in ordinary literature.
2. Nearness to nature.
3. Universal appeal.

This can be quickly disputed – self-consciousness is a defining feaure of particular kinds of literature only and folk stories are certainly conscious of their own conventions; there is no evidence that literature from one folk culture appeals to all people everywhere (or even, as may be implied, in all other folk cultures); and the notion of 'nearness to nature', as anthropologists like Mary Douglas have shown, involves us in not seeing how highly wrought is the sense tribal societies make of the natural world. However, what is important here is that these kinds of ideas about folk culture were believed and inform the development of black images and of Robeson in particular.

This is not to deny that there might be such a thing as folk culture, if we define this simply as those cultures produced in rural, peasant societies – but there is no a priori similarity between those societies and their cultures, and certainly not of the kind evoked by Fausset. That kind of image of folk culture derives most immediately from the Romantic movement (although with a still larger history – see Raymond Williams' **The Country and the City**), and, in the case of its application to black American culture, more specifically from the nationalist schools of music and literature that developed in Europe towards the end of the nineteenth century. The connection between these movements and the kind of folk work Robeson was involved in is quite precise. In the case of music, the link is Anton Dvorak, one of the major examples of a romantic nationalist composer, who used the peasant music of his country as a source and inspiration for his own music. He established, with Jeanette Thurber, the National Conservatory of Music, with the remarkable policy of recruiting equal numbers of white, black and Native American students, in order to foster 'authentic' American music rooted in folk music traditions (see Cruse, 1978, p.54). One of his students was Harry T. Burleigh, whose arrangement of *Deep River,* published in 1916, established the spiritual as a form of art song, acceptable in the concert hall. The spirituals were the cornerstone of the

Harlem Renaissance argument for black folk culture, and were the basis of Robeson's concert and recording work.

The drama connection is similarly precise. The first plays to deal with black subjects from a folk perspective were written by the white Southern playwright, Ridgley Torrence, and he explicitly modelled himself on the work of Yeats, Synge and Lady Gregory in Ireland, probably inspired by the visit of the Abbey Players to New York in 1911 (see Clum, 1969). The connection was seen straightaway by one of the most enthusiastic white supporters of the Harlem Renaissance, Carl Van Vechten – of Torrence's 'Negro play' **Granny Maumee** he wrote 'the whole thing is as real, as fresh, as the beginning of the Irish theatre movement must have been in Dublin' (**New York Press,** 31.3.14, p.12; quoted by Clum, 1969, p.99). Paul Robeson's first stage appearance was in a production of Torrence's **Simon the Cyrenian** (at the Harlem YMCA in 1920), and it was his performance in this that impressed Eugene O'Neill, himself developing a folk-based drama. Robeson subsequently appeared in three of his plays, including **The Emperor Jones** – one of the roles that established his reputation – in the theatre and in the film version. Further developments of this kind of folk drama include **Porgy** by Dorothy and DuBose Heyward (described by James Weldon Johnson (1968, p.211), the first major chronicler of the Harlem Renaissance, as 'a folk play . . . [which] carried conviction through its sincere simplicity'), in a 1927 revival of which Robeson played Crown; and **Porgy and Bess** (1935), a 'folk opera' based on **Porgy,** several of whose songs Robeson recorded with considerable success.

The nationalist movements in music and literature occurred in Europe in those countries at the point of undergoing a rapid spurt of industrialisation and urbanisation, and at a time when the rural, peasant experience that is the basis of folk art was becoming increasingly untypical of the population as a whole. Yet it is just this untypical situation and experience which is labelled, by nationalist aesthetic movements, as the characterising culture of the country. Similarly, the spirituals as concert music and the plays of Torrence, O'Neill, the Heywards and others occur as a celebration of the Southern Negro situation and experience at the end of a massive migration of black people in the USA to the cities of the North, and supremely to Harlem, the urban heartland of these productions of the Southern Negro essence.

The Negro folk idea, and Robeson's relation to it, is probably best looked at through the spirituals rather than the folk dramas. Not only were the spirituals the art form that the Harlem

Renaissance pinned its colours to in terms of its claims for Negro culture, they are also the most enduring feature of Robeson's career, not only in concerts but as a regular part of stage and film appearances. (The films do not necessarily include true spirituals, but they do feature folk material or imitation folk material – and though the issue of genuineness and authenticity can be crucial in discussions of folk, at the level of popular representation the folk feel can pass muster.) There was general agreement that Robeson's approach to the singing of spirituals was 'simple' and 'pure', but the kinds of emotions and meanings that he conveyed through the spirituals were more widely interpreted.

Robeson's musical qualities seem to exemplify Albert C. Barnes' view of the essence of the spiritual – 'natural, naive, untutored, spontaneous' (Locke, 1968, p.21). Mary White Ovington (1927, p.213), a (white) member of the Niagara group and founder member of the NAACP, stressed that Robeson retained the simple essence of the spiritual because he was in touch with its roots in folk culture:

> *Robeson had heard these songs in his boyhood as the older generation had given them, simply, without diminuendo tailpieces or a conductor's pounding of time at each new line . . . Thus Robeson came to his . . . triumph as a singer of spirituals.*

Musically there is no disputing Robeson's simplicity and purity, if we are careful what we mean by these terms.

By simple is meant that he sings the melody straight through, with little adornment whether of the classical singer's trills, slides and other decorations or the jazz use of syncopation or phrasing that bends, delays, quickens or plays on the melody. By pure is meant that his voice always remains within the strict tonal system of Western harmony, not using any of the 'dirty' notes of black blues, gospel and soul music. You are made aware of Robeson's pure and simple approach when you listen to his singing songs that are not written to be sung like that. His recordings of *Summertime* and *A Woman is a Sometime Thing* from **Porgy and Bess** with the Carroll Gibbons Orchestra illustrate this. One does not expect the kind of intensified jazz play of Lena Horne's 1957 recording of *Summertime* or Leontyne Price's syncopated coloratura treatment in her 1981 recording, but even the relatively straightforward recordings and the score itself indicate a number of syncopated and 'impure' elements that Robeson eschews. Where Gershwin has slides and uneven notes on 'Don't you cry' (e.g. 1a), Robeson sings precisely separated,

*Musical examples are given in the clef and key of the original score for ease of comparison; i.e. my renderings of Robeson's versions are not transposed to his register.

even notes (e.g. 1b),* and the syncopation required for phrases like 'an' yo' ma is good-lookin'' (e.g. 2a) become evenly paced notes in Robeson (e.g.2b). These 'purifications' of Gershwin are even more noticeable in *A Woman is a Sometime Thing*, anyway a more jazz related number. All singers that I have heard, save Robeson, bring out the rhythm of a phrase like ''Fore you start a-travellin'' (e.g. 3) by a syncopated stress on ''Fore' and a longer stress on 'tra-' and consequently shorter ones on '-ellin'', details that the classically based notation system cannot render. Robeson, however, gives little stress to ''Fore' and sings 'travellin'' as three notes of correct (as notated) length.

Similarly his only blues recording, of **King Joe (Joe Louis Blues)** with the Count Basie band in 1941, shows, by the concessions it makes to blues style, how little touched by these black music traditions Robeson generally is. At the end of phrases he gets that sense of falling on to a note, characteristic of blues singing, of which perhaps the most memorable, because most lyrically appropriate, example is 'I hate to see the evenin' sun go down'. Thus in **King Joe** he sings:

They say Joe don't talk much
He talks all the **time**
They say Joe don't talk much
He talks all the **time**

He also sings the blue notes of the tune (as, of course, he has to if he is going to sing it at all). But on an exultant final phrase like

But the best is Harlem
When a Joe Louis fight is through

he is too concerned to get all the words in to allow for any hollering or rasping, or even for extending 'fight' briefly but tellingly beyond the strict time-value that the note accords it.

If Robeson sings without any of the conventions of jazz blues, his form of folk singing also bears little resemblance to the kind of nasal delivery that is now customary in folk and folk-inspired music. Robeson's singing is art or concert singing without the flourishes – hence pure of tone and simple of delivery. And this purity and simplicity was taken to be the hallmark of singing which caught the folk essence of the spirituals. Robeson was held to return to the true basis of the spiritual, without recourse to symphonic elaboration, as Ovington noted above, and without, as Elizabeth Sergeant (1926, p.207) put it, 'lapses into jazzed effects and the Russian harmonies that have recently crept into the Spirituals'.

If symphonic and jazz inflections were not true to the spiritual, then neither was pure and simple singing either, if by true is meant accurately reproducing how they were originally sung. The development of the spiritual as a form of concert music was a history of the purification of the spirituals of all their 'dirtiness', intricacy and complexity. To begin with, they were not solo songs anyway but choral, often using elaborate forms of harmony and counter-melody, and the call-and-response patterns characteristic of African music. Polyphony was achieved through the use of vocal timbres, and this subtle sound colouring was developed through the use of various kinds of instrumental accompaniment. Finally, they were rhythmically very complex, making great use of contrapuntal elements (see Roach, 1973, pp.31–3). None of this survives into the concert hall, or record-selling, spiritual. Indeed such elements were consciously expunged both by Harry T. Burleigh (see Southern, 1971a, pp.286–7) and by George L. White, choirmaster of the Fisk Jubilee Singers, the first choir to popularise spirituals in the USA. White aimed for 'Finish, precision and sincerity' (quoted by Roberts, 1973, p.161); in effect, purity and simplicity.

It may well be true that, had it not been for these adjustments, this black music tradition might never have become as widely popular, and marketable, as it did. But purity and simplicity are not only neutral musical terms, they are also value terms, they describe certain ethical qualities that folk cultures are held to maintain and that are supposed to be expressed in folk music. Musical purity and simplicity become purity and simplicity of heart.

The idea that the Negro folk character is itself simple and pure runs through much of **The New Negro** and texts inspired by it. Many of these texts are quite nuanced. Locke (1968, p.200), for instance, stresses that simplicity of means does not imply simplicity of meaning:

> *For what general opinion regards as simple and transparent about [the spirituals] is in technical ways, though instinctive, very intricate and complex, and what is taken as whimsical and child-like is in truth, though naive, very profound.*

Moreover, in the black discourse the spirituals are seen – though not invariably – as cultural products rather than an innate predisposition granted along with dark skin.

These nuances and culture-based conceptions of simplicity and purity were easily lost sight of in white discourses, probably under pressure from what remained one of the most powerful images of blackness, Uncle Tom.* Harriet Beecher Stowe's novel fixed a particular conception of the black character in liberal white discourse, and it is one founded on notions of purity and simplicity. Her description of Uncle Tom could be a description of Paul Robeson (or one way of seeing him), not only because of the character traits but because of the physical similarity:

*The epithet Unle Tom has become so associated with black accommodationist, selling-out positions that the nature of its earlier force as an image is sometimes forgotten. In raising it here in connection with Robeson I am not passing judgement on whether he was an Uncle Tom in this later political sense.

> *a large, broad-chested, powerfully-made man, of a full glossy black, and a face whose truly African features were characterised by an expression of grave and steady good sense, united with much kindliness and benevolence. There was something about his whole air, self-respecting and dignified, yet united with a humble and confiding simplicity* (Stowe, 1981, p.68).

Size, dignity, simplicity – these terms recur so frequently with reference to Robeson that you will find them scattered throughout the quotations used in this chapter. There were, besides, more or less direct references to Robeson as Uncle Tom throughout his career. The **New York Times** wrote of him in **All God's Chillun Got Wings,** 'the hero . . . is as admirable, honest and loyal as Uncle Tom' (quoted in Seton, 1958, p.64), and the play itself ends with the defeat of Jim's plans to become a lawyer and his acceptance of playing Ella's 'Uncle Jim'. *Old Man River* was often seen in Uncle Tom-ish terms – Brooks Atkinson greeted Robeson's singing of it in the New York revival as an expression of 'the humble patience of the Negro race' (20.5.32; quoted by Schlosser, 1970, p.142), and Robert Garland referred to Robeson's Joe as 'one of God's children with a song in his soul' (**New York World Telegram,** 20.5.32; quoted by Schlosser, ibid.). Similar interpretations met his film performance, although for the black press this was a negative quality:

> *if indeed he has ever died in the theater or on the screen, Uncle Tom has a true exponent in Paul Robeson* (**California News,** 8.3.36; quoted by Schlosser, ibid., p.252).

I have already quoted Alan Brien's reference to Robeson's 1959 Othello as Uncle Tom.

Stowe, like other white people later – compare Mabel Dodge's words quoted above – developed an image of blackness as a repository of all the qualities that she considered lacking in the dominant society of her day. As George M. Frederickson (1972) shows in his **The Black Image in the White Mind, Uncle Tom's Cabin** is not only a protest against slavery but a wider critique of contemporary society as materialistic, aggressive and over rational. Stowe sets many of her black characters, and also many of her female ones, against this image of society. Her argument – and hers is only the most imaginatively powerful expression of widespread liberal thought – was that, since blacks were so pure and simple, it was unchristian to treat them badly and that, in their purity and simplicity, they held a lesson for white society. But it was possible for her listeners not to hear the second part of the argument and to draw a different conclusion from her compelling image of the negro character. Her imagery is based on earlier pro-slavery literature that had depicted the Negro as docile and humble, as long as he or she remained a slave; that argued that blacks had a predisposition to bondage, and that it was against their natures to be free; that blacks were happiest kept in their lowly pure and simple state. Moreover, in validating blacks' lack of aggression, Stowe offered no model of black advancement by any means, violent or otherwise. If blacks are so wonderful oppressed, what need have they of liberation? Indeed, are they in that case really oppressed at all?

This is the crux of the Uncle Tom image. By validating Negro qualities it keeps the Negro in her/his place, and it permits the dimension of oppression, of slavery and racism, to be written out of the white perception of the black experience. The simple and pure distillation of the Negro essence in the development of the spirituals and Robeson's singing of them purges blackness of the scars of slavery, of the recognition of racism. At most these elements are marginalised or rendered invisible. None of Robeson's films confronts slavery or racism directly, but neither was it in the meaning of the spirituals and Robeson's singing of them. Or rather – and this is the point – it was possible not to hear it there, not to discern it in his much vaunted 'expressiveness'.

To say that Robeson sang simple and pure does not mean that he sang without expression. He had a deep, sonorous voice, but was able to soften it to convey a range of gentle, tender, sweet, melancholy qualities. Equally he used not so much loudness as intensification of his voice to convey strong, deeply felt emotion. In his rendition of *Deep River* in **The Proud Valley** (1939), the close-up allows us to see the effort involved in the singing, the deep breaths, the moistening of the lips between phrases, which signal the intensity of the feeling that is being produced. Softening and intensification of a rich, deep voice – these means were constantly found profoundly, movingly expressive; but expressive of what?

The answer, or rather answers, to that are suggested if we consider what the spirituals were held to express. This is a subject of controversy. More clearly here than in the general discussions of the negro folk character, one can see the different way white and black discourses hear and feel the same material.

One might say that the ultimate white person's spiritual is *Old Man River,* though of course it is not a spiritual at all. But it expresses the basic emotional tone that whites heard in spirituals and in Robeson's voice – sorrowing, melancholy, suffering. Moreover, the way it reiterates the river imagery allows the cause of this suffering to be laid at the door of the river's indifference and to be transmitted into the eternal lot of mankind, as borne, conveniently for the white half of the community, by blacks. *Old Man River* does contain references to slavery and oppression, even before Robeson started changing the lyrics in an effort of resistance against the song's more generally heard message of resignation. There is reference to the fact that 'darkies all work while de white man play' and 'you don't dast make de white boss frown' but this is swallowed up in the generalised reference to 'you and me' who 'sweat and strain' in the face of the endlessly rolling river.

It is this mood of sorrow and resignation that whites

heard, after all without difficulty, in Robeson favourites like *Nobody Knows the Trouble I've Seen* and *Sometimes I Feel like a Motherless Child.* River imagery, apparently echoing the melancholy of *Old Man River,* recurs again and again in his song repertoire – *Deep River, Swing Low Sweet Chariot* ('I looked over Jordan . . .'), *River Stay way from my Door, Sleepy River* written for **Song of Freedom,** *The Volga Boat Song,* and so on.

River imagery is very common in black American folk music – Jerome Kern and Oscar Hammerstein were right to put it into the mouth of the chief Negro character in **Show Boat,** but wrong to make it mean the eternal vale of life's tears. Within the context of slavery, river imagery has a dense set of meanings, all to do with deliverance from the oppression of slavery. It could refer to hope in the hereafter, crossing over the river of death into heaven, but it could also refer to rivers of escape, to the North and Canada, or else to the Atlantic Ocean, crossing that river not only to escape but to return to the homeland, Africa. The familiar spirituals (though not, of course, *Old Man River* or *Sleepy River* that were composed in line with the melancholy-resignation image of the spirituals) take on a very different significance when heard this way. *Sometimes I Feel like a Motherless Child* is not just an ever-sorrowing lament but a reference to the fact of slavery 'a long ways from home' (i.e. Africa), and the second verse, 'Sometimes I feel like I'm almost gone', is no longer a mournful statement of weariness and pain, but a recognition of death or escape as a release from slavery. The fourth line of the refrain of *Deep River,* 'I want to cross over into camp ground', is meaningless to white ears but a reference in black tradition to the secret camp meetings where it was possible to meet slaves newly arrived from Africa and hence have contact with the homeland (see Southern, 1971b, pp.82–7, and Walton, 1972, pp.25–8). In this reading, the spirituals are not about some innate Negro predisposition to suffering, but about the suffering inflicted on generations of black people by slavery and, by extension, by white racism since abolition.

But slavery could not be spoken – it represented too much guilt for whites, too much shame for blacks. Thus even though in interviews Robeson often stressed that he thought he was expressing the sufferings of slavery and the slaves' hope for release, it was easy to hear those deep, gentle, intensified cadences as expressive of a more generalised racial feeling. Slavery and racism would base the black folk image in historical reality; but if you didn't hear that element in the spirituals, see it in Robeson, the folk image became the unchanging, unchangeable fate of the black people.

Atavism

The idea of the black race as a repository of uncontaminated feeling informs the atavistic image as much as the folk image, and both have been acceptable to blacks and to whites. But whereas the folk image can be read as an Uncle Tom image, the atavistic image can be read through the other major male black stereotype, the Brute or Beast.

Atavism as a term need only mean the recovery of qualities and values held by one's ancestors, but it carries with it rather more vivid connotations. It implies the recovery of qualities that have been carried in the blood from generation to generation, and this may also be crossed with a certain kind of Freudianism, suggesting repressed or taboo impulses and emotions that may be recovered. It also suggests raw, violent, chaotic and 'primitive' emotions. It is this complex of ideas and feelings that are at issue when considering notions of atavism in relation to blackness and Robeson.

Atavism, in the black American/Robeson context, involves first the question of Africa, for it was to Africa that the unrepressed black psyche was held to return. What Africa was presumed to be like was then also what blacks were supposed to be like deep down, and the potent atavistic image of Africa acted as a guarantee of the authentic wildness within of the people who had come from there. This was held equally, though with different understandings, within black and white discourses – at one level, atavism was simply the tougher, more energetic version of the folk image. Robeson's relation to both the African and the more generalised atavistic image is quite complicated. In principle embracing Africa as a homeland, his film and stage work is implicitly a rejection of the idea of Africa; and while initially associated with the more general wildness-within idea of atavism, this is dropped quite early on in his career, too close to the brute stereotype to appease white fears or please black aspirations.

Africa

An initial problem was that of knowing what Africa was like. There is an emphasis in much of the work Robeson is associated with on being authentic. The tendency is to assume that if you have an actual African doing something, or use actual African languages or dance movements, you will capture the truly African. In the African dream section of **Taboo** (1922), the first professional stage play Robeson was in, there was 'an African dance done by C. Kamba Simargo, a native' (Johnson, 1968, p.192); for **Basalik** (1935), 'real' African dancers were employed (Schlosser, 1970, p.156). The titles for **The Emperor Jones** (1933) tell us that the tom-toms have been 'anthropologically recorded', and several of the films use ethnographic props and footage – **Sanders of the River** (1934, conical huts, kraals, canoes, shields, calabashes and spears, cf. Schlosser, 1970, p.234), **Song of Freedom** (1936, Devil Dancers of Sierra Leone, cf. ibid., p.256) and **King Solomon's Mines** (1936). Princess Gaza in **Jericho** (1937) is played by the real life African princess Kouka of Sudan. Robeson was also widely known to have researched a great deal into African culture; his concerts often included brief lectures demonstrating the similarity between the structures of African folk song and that of other, both Western and Eastern, cultures (see Schlosser, 1970, p.332). However, this authentication of the African elements in his work is beset with problems. In practice, these are genuine notes inserted into works produced decidedly within American and British discourses on Africa. These moments of song, dance, speech and stage presence are either inflected by the containing discourses as Savage Africa or else remain opaque, folkloric, touristic. No doubt the ethnographic footage of dances in the British films records complex ritual meanings, but the films give us no idea what these are and so they remain mysterious savagery. Moreover, as is discussed later, Robeson himself is for the most part distinguished from these elements rather than identified with them; they remain 'other'. This authentication enterprise also falls foul of being only empirically authentic – it lacks a concern with the paradigms through which one observes any empirical phenomenon. Not only are the 'real' African elements left undefended from their immediate theatrical or filmic context, they have already been perceived through discourses on Africa that have labelled them primitive, often with a flattering intention.

This is not just a question of white, racist views of Africa. It springs from the problem, as Marion Berghahn (1977) notes, that black American knowledge of Africa also comes largely through white sources. It has to come to terms with the image of Africa in those sources, and very often in picking out for rejection the obvious racism there is a tendency to assume that what is left over is a residue of transparent knowledge about Africa. To put the problem more directly, and with an echo of DuBois' notion of the 'twoness' of the black American – when confronting Africa, the black Westerner has to cope with the fact that she or he is of the West. The problem and its sometimes bitter ironies, is

At the Giza pyramids in Egypt, with Wallace Ford and Henry Wilcoxon during filming of **Jericho** (1937)

illustrated in two publicity photos from Robeson films. The first, from the later film **Jericho,** shows Robeson with Wallace Ford and Henry Wilcoxon during the filming in 1937. It is a classic tourist photo, friends snapped before a famous landmark. Robeson is dressed in Western clothes, and grouped between the two white men; they are even, by chance no doubt, grouped at a break in the row of palm trees behind them. They are not part of the landscape, they are visiting it.

The second photo is of Robeson and Jomo Kenyatta taken during the shooting of **Sanders of the River.** Kenyatta, the African and future African leader who plays a chief in the film, wears white Western male attire; Robeson, the American who plays the part of a leader in all his films set in Africa, wears the standard Hollywood native (or Tarzan) get-up. As Ted Polhemus and Lynn Proctor (1978, p.93) point out, clothes are major signifiers of power on the international political scene. When the West is in the ascendant, other nations dress in Western clothes; but when the relations of power shift, the leaders of non-Western nations can wear their national clothes. In the Robeson-Kenyatta photo, the politician (as he was then learning to be) who would have real power in Africa implicitly acknowledges Western cultural power in his dress; while the actor who enacts an idea of power in Africa both explicitly rejects the West in his dress and yet in fact asserts Western notions since the dress owes far more to Western ideas about African dress than African ideas of it. It is hard to think of a more graphic example of how much a black American star is inevitably caught in Western discourses of

On location with Jomo Kenyatta for **Sanders of the River** (1935)

blackness, especially when seeking blackness in African culture rather than in black American culture itself.

Something of this problem is confronted in **Song of Freedom.** Here, unconsciously perhaps, is an engagement with the problem for black Americans of coming to terms with Africa. When Robeson/John Zinga arrives on Casanga (the island whose rightful monarch he is, being descended from the despotic queen Zinga who had reigned a couple of centuries earlier), he is rejected by the inhabitants because of his white man's clothes. 'Everything's so different from what I expected – it's all so primitive,' John says. He further alienates himself by breaking up a witch doctor's dance, since he believes his white man's medicine will do more good for a sick person than superstitious African dancing. He has to learn to show respect for the customs of Casanga, and it is only when he reveals that he knows the ceremonial song that 'no white man has ever heard' that he is finally accepted as the true king of the island.

The film tries to bring into line a progressive black approach to Africa and an atavistic image of Africa. John's ideas about medicine are part of the conception in black discourses of a synthesis between black spirituality and white technology; but there is no image of black African spirituality in the film, only mumbo jumbo. The positive images of African spirituality in the literary work of Hurston, McKay and others created no equivalents in popular culture; John/Robeson's spoken ideas of black and white synthesis founders on the lack of any visualisation of what the positive black contribution to that synthesis would be. The same kind of contradiction dogs, though less damagingly, the climax of the film, with John's remembering the words of the *Song of Freedom*. The fact that it is a song of freedom links it to a central fact of black experience, a point to which I'll return; and its narrative function is to bring order and stability to Casanga (under the rightful hereditary leader to the throne, which John is), not primeval chaos. However, the fact that John has carried a snatch

Voodoo (1922): Robeson as 'handsome, genial plantation worker' . . .

. . . and as 'manic African savage'

of the song in his head throughout the film and that it all comes back to him at the climax suggests the (atavistic) idea of a race memory, and this is realised musically through a 'jungle' beat and words that suggest a release from the savagery of thunder, wild bird and lion, climaxing with the Western John Zinga/Paul Robeson singing

From the shadow of darkness
I lead my people to freedom.

Song of Freedom is in tension with the atavistic idea of Africa, which was central to two earlier stage works, **Voodoo** and **The Emperor Jones.** Unfortunately it is hard to find out a great deal about **Voodoo*** but it did involve the Robeson character, a plantation worker (pre or post slavery?), falling asleep and in his sleep returning to Africa. Photos of the production (which look like front-of-house pictures rather than photos of the play in performance) clearly show Robeson as a handsome, genial plantation worker and then as a manic African savage. In his dream, the plantation worker has relived his ancestry as a wild man. The idea that seems to be implicit here, that within the civilised black man there is the trace of savage ancestry, is more explicitly and definitely present in **The Emperor Jones** as both play and film. Brutus Jones is a Southerner who, through cunning and threat, comes to be boss man and self-styled 'emperor' of a Caribbean island; but when his rule becomes intolerable, the

*Besides – Marie Seton (1958, p.58): 'The central point of the play was never clear.'

islanders rebel and chase him through the forests of the island. The image of Jones in flight takes up nearly all the play and the larger part of the film (which includes more about Jones's life in the USA), and it is the image that is fixed as the core of the play.**

Song of Freedom, in so many ways a vehicle for Robeson, includes a performance by John Zinga of *The Black Emperor*, an opera that is obviously a reference to **The Emperor Jones even down to the similarity of the costume; we see only the emperor in the forest in his agony of flight – the scene is virtually meaningless if the viewer is not familiar with **The Emperor Jones,** but equally the scene is wholly sufficient to evoke the received idea of the play.

The significance of this long drawn out flight is that in the course of it Jones reverts from being a civilised ruler to being a chaotic, superstitious primitive. He becomes primeval, aboriginal. He stumbles deeper into the forest (+ jungle + Africa + chaos); he has hallucinations of wild beasts (alligators) and of witch doctors, as well as of moments from his US past. His manner and language become more crazed and fearful. It is a classic statement of the atavistic idea in the black context – that blacks have within them, beneath a veneer of civilisation, a chaos of primeval emotions that under stress will return, the raging repressed.

This return of the primeval repressed does not recur in any of Robeson's later roles, but the image of Africa in several of his British films does evoke these same primitivist ideas. This is not done, however, through the Robeson character himself, but rather through the general construction of an image of Africa. Perhaps the most offensively sustained example of this is **Sanders of the River.**

Throughout the film Africa is represented as a land of at best childish incompetence and at worst dangerous, violent ferment. The message of the film is that it is only the presence of white men that brings any law and order to the continent. The most concise and vivid illustration of this are two montage sequences, one occurring when Sanders has left the area (to go to Britain to get married) and the other on his return. The first showing Africa reverting to chaos – the drums of the 'bad' tribes sending messages that Sanders is dead, 'there is no law any more' are cut with shots of animal wildlife, which then cut to dissolving shots of war dances and warriors marching, followed by a series of dissolves of shots of vultures, ending with more shots of drumming and marching. The sequence dramatically shows the absence of the good white ruler heralding a reversion to a state of nature which, in Africa, is also a state of chaos and threatening violence. The second sequence reverses this. Shots of the plane carrying Sanders back to the area are cut with further wildlife footage, but whereas in the earlier sequence the animals are for the most part the dangerous ones (lions, crocodiles), here they are the more harmless ones (birds, wildebeest), and where earlier they were crowding together or moving in for the kill, here they are scattering. As these latter images are also aerial shots, this gives the impression of the animals of Africa scattering before the return of the white boss in his plane. This is further linked to the end of the immediately previous sequence, where the ritual war dance of the rebellious tribes ends in a charge accompanied by catcalling. This charge is executed *towards* the camera, down left of the frame. The movement of the aerial camera in the Sanders return montage is forward into the space being filmed, and the animals scatter to left and right *away from* the camera. We are thus placed in a position that is first fearful of, and then in fantasy can quell, primitive revolt.

In the two films in which he plays an African (**Sanders of the River, King Solomon's Mines**), Robeson is not associated with this primitivism, but is rather presented as the exceptional figure. He is the 'good' African with whom the whites can work. There is a difference of emphasis between the two films – in **Sanders of the River** Bosambo is dependent on Sanders both for his authority (Sanders sanctions his leadership which means that his rule has white power as a back-up) and directly in the plot when he and Lilonga have to be rescued by Sanders from captivity. In **King Solomon's Mines** there is more sense of the whites being dependent on Umbopa, who knows the desert, is able to smell the whereabouts of water and uses his massive strength to push a boulder away from the entrance to the mines where the party are trapped. Yet Umbopa is, like Bosambo, dependent on white intervention to establish his rule – it is by using white 'magic' (knowledge of eclipses) that Umbopa plays upon the superstitions of 'his people' to convince them he is their rightful leader.

Broadly then, in Robeson's African roles he is, at the level of narrative structure, set apart from African primitive violence. This is largely true too of other aspects of the way the character is constructed. His singing, a selling point of the films, is markedly different from other singing in the films, though both are supposed to be African. Robeson/Bosambo leads the just (white-backed) war against the usurping king Mofolaba with a steady marching song, with English words in the tradition of European/American battle songs, and delivered in his booming deep voice. Later in the film, when Ferguson, Sanders' replacement, has been captured, he is surrounded by men in masks, singing in cod African, to a hectic, irregular beat, in high-pitched voices – in every particular the antithesis of the Bosambo/Robeson war song. Similarly in **King Solomon's Mines** Umbopa/Robeson sings English lyrics to melodies with a regular (unsyncopated, uncomplex) rhythm and Western harmonies, where 'his people' sing African words and music. The latter, as with much of the footage in both **Sanders of the River** and **King Solomon's Mines,** is 'authentic', in the sense of being recorded on location; when, during the eclipse, Umbopa/Robeson sings in African, it is almost certainly Efik, which he learnt for the part. But the placing of these 'genuine' African elements is significant. In the ceremonial dance done near the mines, at which Gagool, the (wicked, in the film's terms) wise woman of the tribe, is to select those to be sacrificed, the African singing and dancing, however authentic, functions in the narrative as a threat to our identification figures, one of whom, Umbopa/Robeson, is himself African. Moreover, an isolated shot from this scene looks like any photograph of white dignitaries looking on at African tribal performance (cf. the Royal Family's obligatory look at native customs on their visits to colonies and former colonies) – the position of 'our' characters in the frame, the way they hold themselves and direct their attention, reproduces the relations of power and difference of colonialism, even if in the plot it is the (to all intents and purposes) white party that is under threat.

The Robeson characters are also written as docile, good, simple characters. It is sometimes argued that within that, at the level of performance, Robeson himself suggests a level of insolence or irony that undercuts this Uncle Tom/Sambo stereotype, but this is a hard argument to sustain. It is not just editing alone (as is also often claimed) that makes Bosambo appear a lovable, rather naughty child in **Sanders of the River;** nor in his first meeting is it just the powerful headmaster–schoolboy rhythms of the script, though these are extraordinarily evocative:

SANDERS *You know that every six months I call on the Orkery.*
BOSAMBO *Yes, Lord, at the time of the taxes.*
SANDERS *Yes, Bosambo, at the time of the taxes, in a month's time.*

Robeson's performance reinforces this characterisation, often quite subtly. When Sanders questions him about whether he is a Christian, he says,

I went to missionary school, I know all about Markee, Lukee, Johnee and that certain Johnee who lost his head over a ransom girl,

but Sanders cuts him short with, 'That will do.' Robeson/Bosambo speaks the speech just quoted with great eagerness, warming to the subject as it becomes more sexual, but then looks utterly crestfallen when Sanders stops him. This is in a two-shot, not a shot/reverse shot sequence; the eagerness and crestfallenness are conveyed through acting.

In all these ways then Robeson, unlike his roles in **Voodoo** and **The Emperor Jones,** is not identified with the atavistic view of Africa as a state of inchoate, dangerous emotion. There are perhaps slight hints of it. Bosambo's thoughts run easily to sex, as seen above, and while Sanders has to leave Africa to go away to marry, a journey that is never consummated since he has to return to Africa to quell the chaos his departure occasions, Bosambo is shown surrounded by adoring women and later marries Lilonga/Nina Mae McKinney, on the strength of an extremely sexually explicit dance she performs. Africans are allowed the sexuality that whites stoically (or relievedly) have to do without. The writing of the film of **King Solomon's Mines** does still have something of the ambivalence of Rider Haggard's novel, where the narrator Allan Quartermaine both admires Umbopa ('I never saw a finer native' (1958, p.42), 'a cheerful savage . . . in a dignified sort of way' (ibid., p.46) and yet is also affronted by his arrogance ('I am not accustomed to be talked to in that way by Kaffirs' (ibid., p.58)). Here too the contrast is with other, inferior, more African Africans – whereas Umbopa is 'magnificent-looking . . . very shapely . . . his skin looked scarcely more than dark' (ibid., p.42), Twala, king of the Kukuanas, is described as 'repulsive . . . the lips were as thick as a Negro's, the nose was flat . . . cruel and sensual to a degree' (ibid., p.115). But hints of sexuality, traces of arrogance are but hints and traces – if anything, it would seem that such violent, difficult qualities have been eliminated, because they are threatening elements. What remains is Robeson's passive physical presence, something I'll come back to later.

Wildness Within

Atavism as a reversion to Africa is then only a feature of a couple of Robeson's stage roles, and of the image of Africa in his films but not of his role in those films. The more generalised idea of atavism as wildness within is implicit in some of the other roles where the connection with Africa is not explicitly made. As an image, it is very close to the alternative black male stereotype to Uncle Tom and Sambo, namely the brute or beast.

This informs Robeson's roles not only in **Voodoo** and **The Emperor Jones,** but in **Black Boy** (1926), Crown in **Porgy** (1928) and in **The Hairy Ape** (1931). Even when this was not clear in the role, there was often a play on its sexual dimensions. Hints of black-white rape cling to the characters – in **Basalik** (1935) Robeson played an African chief who abducts (but does not in fact rape) the wife of a corrupt white governor; in **Stevedore** (1935), he played Lonnie Thompson, who is wrongly accused of raping a white woman at the start of the play. Equally, **Othello** could be read in this light – Robeson could not take the 1930 London production to Broadway because it was felt improper for an actual black man to play Othello to an actual white actress. When **All God's Chillun Got Wings** had been produced in New York in 1924 an enormous amount of publicity had surrounded its depiction of a black–white marriage even before its first performance. Although not the first play to deal with this topic, a perennial theme in American literature, a forceful Robeson/O'Neill/Provincetown Players treatment was anticipated as dangerous and subversive. Even when films Robeson appeared in were not in the brute mould, the advertising for them might hint that they were – **Sanders of the River** was sold with the following come-on: 'A million mad savages fighting for one beautiful woman! Until three white comrades ALONE pitched into the fray and quelled the bloody revolt' (quoted by Cripps, 1977, p.316).

The brute image does not only have a sexual dimension. As a football player, though praised largely in terms of skill and intelligence, his playing was also described as 'vicious' – 'no less

Robeson the football player

than three Fordham men were sent into the game at different times to take the place of those who had been battered and bruised by Robeson' (**New York Times,** 28.10.17; quoted by Henderson, 1939, p.100). Writings about him on the field are a litany of references to his size – 'the giant Negro', 'the big Rutgers Negro' (ibid.), 'a dark cloud . . . Robeson, the giant Negro' (**New York Tribune,** 28.10.17, Charles A. Taylor; quoted by Hamilton, 1974, p.29), 'a veritable Othello of battle . . . a grim, silent and compelling figure' (quoted without reference by Ovington, 1927, p.207); and publicity for his appearances suggests a force and dynamism that could have been threatening.

The brute stereotype not only threatened whites, it also offended blacks. The fact that it was abandoned, along with one of its justifications, atavism, early on in Robeson's career, is a result of common cause between the two ethnic discourses. But there is a further suppressed element – slavery, again. Just as the spirituals were emptied of their slavery meanings and transformed into songs of universal suffering, so one of the possible articulations of atavism was also lost. The primary meaning of atavism was back-to-the-aboriginal-jungle, but it was possible to return to another violent race memory, that of the experience of slavery. The play of **The Emperor Jones** does just this. Jones's flight into the forest is systematically organised as a flight through Afro-American history – after hallucinating a general, Expressionistic vision of 'little formless fears' ('if they have any describable form at all it is that of a grubworm about the size of a creeping child', Scene Two), in the following scenes Jones sees, in this order, a Pullman porter playing dice, a black chain gang with a white guard, a group of white slave buyers 'dressed in Southern costumes of the period of the fifties of the last century' (Scene Five), two rows of Negro slaves huddled together in a slave ship, a Congo witch doctor.

Each of these visions is emblematic of black American history – Pullman porters were an all black trade and probably the most familiar black figures in American public life, as well as a respectable avenue of employment for black men; the chain gang was not an exclusively black experience but, then as now, blacks were disproportionately represented in penal institutions, as all oppressed peoples are; the slave trade and, finally, or rather as point of origin, Africa need no further gloss as aspects of black history. O'Neill's use of the witch doctor clearly brings his image of Africa in line with that discussed above, and the idea of this Africanness being in all black people still is brought out in the last scene. Throughout Jones's flight tom-toms have been beaten, becoming gradually louder and faster; in the final scene, with Jones now dead, Smithers, a white trader who was in league with Jones as ruler of the island, says to the islanders who have been pursuing Jones,

I tole yer yer'd lose 'im, didn't I? – wastin' the 'ole bloomin' night beatin' yer bloody drum and castin' yer silly spells!

The dramatic irony is not only that we know that Jones is dead, but that his terror has been shown on stage as induced by or at least whipped up by the relentless tom-toms and his death comes as he fires his gun wildly at his final hallucination, the witch doctor. Smithers is doubly wrong – Jones is dead, and it was the drums and spells that got him. The conclusion of the play is then African atavism, but the structure of the rest of the play emphasises Jones's terror at the real, historical experience of the black man in America.

In the film, the sequence of visions is altered. Most significantly, there is no reference to slavery at all. The progression of visions seen in flight are – Jeff, a Pullman porter whom Jones worked with, playing dice; the chain gang; a Southern black church service; the witch doctor and an alligator. One of the effects of these changes is to individualise Jones's flight into his (race) memory, partly because the first three visions use shots from earlier in the film and thus retrace Jones's career before arriving on the island. Moreover, because the film has given us more of the narrative before Jones's arrival on the island, we know that it is a brawl arising out of a dice game that led to his conviction – the chain gang becomes less emblematic of a condition of oppression, and more a moment in a particular person's biography.

This individualisation goes hand-in-hand with a fundamental alteration of the facts and meaning of black history. Whereas the play gives us

low-life black employment → the ordeal of the black chain gang → the terrors of slavery → the fears of mumbo-jumbo

the film gives us

low-life black employment → the ordeal of the chain gang → superstitious black Christianity → superstitious mumbo-jumbo

Neither progression is entirely 'logical', but the difference of emphasis is striking. Moreover, although there is a limit to what one can say about the production of **The Emperor Jones** as a play (even given O'Neill's quite detailed stage directions, and even if one confined onself to the Robeson productions), the film does seem to emphasise inchoate emotion over the terrors of history. There is a visual progression in the film as Jones gets further and further into the jungle – that is, his primitive self. The trees, creepers and forest shadows crowd out the screen more and more, so that we see Jones *through* the jungle. There is no indication in the play of any equivalent devices for the theatre. In the film, when Jones sees the chain gang, he joins in, beating with an imaginary hammer before his anger leads him to shoot at the white guard. The play here emphasises white violence – Jones's action is provoked by the guard's cruelty with the whip that he uses to lash both the gang in general and Jones in particular. Lastly, the film opens with a shot of African tom-toms being beaten which dissolves to jazz drumming; nowhere does the play insist on a direct continuity of African and Afro-American feeling. In such ways, the film empties the play of its historical dimension, of the reality of black oppression.

Robeson's relation to the atavistic image of blacks was blocked on two counts. The possible inflection of it in terms of slavery was dropped – white guilt and black shame made it too uncomfortable an element for cross-over stardom. But without this, atavism is too close to the brute and beast image, too threatening to whites, too demeaning for blacks. It was thus eliminated gradually from his work, as a disturbing undertone that could either be ignored or seen as something that is contained or withheld.

Freedom

They are scared, Negro brother,
Our songs scare them, Robeson.
(Nazim Hikmet, 'To Paul Robeson')

So far I have been arguing that the conditions that made it possible for Robeson to be a cross-over star were his place in discourses of blackness that were acceptable to both black and white audiences, though they might be differently understood by those audiences. There is a price to be paid for success in these terms, and I have indicated how easily a positive view of black folk and African culture as a radical alternative to materialistic, rationalistic, alienated white Western culture slides back into sambo, Uncle Tom, brute and beast.

It is important to emphasise what nonetheless remains in the Robeson image, even when the sambo or brute imagery dominated it: a certain irreducible core of meanings that were still felt to be threatening or insolent or inspiring enough to need further ideological handling.

To begin with, a major emphasis of the image was always Robeson's physique, and the size and muscularity of this could easily be seen to be potentially at the service of the strength, power and action necessary for radical black advancement. Equally, his unassailable reputation as good, noble, pure, and so on, however much we may note in it notions of naivety and simplicity (and behind them childlikeness), were an image of what a black person could be and was, an image that still had that moral force in the face of racism which Stowe and others had intended it should have. Moreover, the extraordinary record of Robeson's achievement, outlined at the beginning of this chapter, meant that it was hard to deny what a black person was capable of doing – in sports, in the professions, academically and artistically.

More important still is the insistence on the concept of freedom throughout his work. I have noted above how the fact of slavery is to a considerable extent suppressed in Robeson's work. Although not altogether – it is still there in the play of **The Emperor Jones**; it is there obliquely in the words to *Old Man River;* it is there in the establishing sequence of **Song of Freedom** where John Zinga's ancestors flee the despotic queen Zinga unwittingly into the arms of the white slave trade.

Yet if direct acknowledgement of slavery is only to be found in these few examples, the implication of it and of white racism in general is continually present through the longing for and quest for freedom that runs through so much of his work. The song John Zinga carries in his (race) memory is not just any old song, but a song of freedom, giving its name to the film. Here, Africa connotes above all freedom, and, as Marion Berghahn (1977) has pointed out, this was characteristic of much black cultural production of the period, though seldom of white. At the beginning of **Jericho** (1937), Jericho/Robeson is unfairly sentenced to death for his necessary act of shooting a panic stricken white sergeant when their ship was torpedoed; he escapes, to the freedom of Africa, and when his friend, the white captain Mack, finds him and tells him he can clear himself, he, Mack, realises that Jericho is in fact happier in Africa with his people than he could be back in the USA. Once again, Africa symbolises the freedom from oppression represented by the USA. Even in the unpromising setting of **My Song Goes Forth** (original title, **Africa**

Looks Up, 1936), a documentary on South Africa, purporting to show what 'the white man has achieved for himself' and 'what he has done for the natives' (**Africa Looks Up** publicity booklet, n.d., p.l; quoted by Schlosser, 1970, p.524), Robeson sings of freedom and Africa:

From African jungle, kraal and hut
Where shadows fall or torrid light,
My song Goes Forth and supplicates
In quest of love and right
I seek that star which far or near,
Shows all mankind a pathway clear,
To do unto his brother
And banish hate and fear.

(**My Song Goes Forth** publicity sheet, Ambassador Film Productions, n.d., p.2; quoted by Schlosser, ibid., p.255.)

In his concerts, Robeson increasingly stressed a similar message, either by virtue of the conditions of performance (e.g. low and uniform seat prices; open-air concerts) and where performed (e.g. at Loyalist camps in Spain in 1938) or by introducing more political material, culminating in his 1942 recordings of **Songs of Free Men,** including *Joe Hill* and Soviet and Spanish Loyalist songs, which he had been using for several years in concerts. By this time his explicit espousal of socialist ideas was widely known, and often used against him. It is not this connection which is crucial here, but the idea of freedom that is important, since this is so fundamental a part of American rhetoric. As Michael Klein (1975) has argued in relation to **Native Land** (1941), the left-wing Frontier Films' documentary for which Robeson did the narration, a possible strategy for left film-makers has been to appropriate the general national ideological rhetoric for left struggle, to demonstrate that freedom is not the prerogative of the right but is endemic to left thought too. In his general appeal to freedom, Robeson can be understood as employing just this strategy.

Parenthetically, we should note the limit of this strategy (as of any other). **Native Land** is notable for having a black narrator, yet a tiny handful of black images. The incident used to exemplify the vileness of the Ku–Klux–Klan is when they attack some white men. At the start of a scene at a union meeting, the camera gives us a close-up of a black male face and then moves across to a white woman sitting next to him; there is then a cut to a series of shots of white men. The black man and white woman are joined together in one binding shot, and then the dominance of white males is reaffirmed. The only black women in the film are those shown briefly dancing towards the end. Apart from the extreme visual marginalisation of black people, there is also something very disconcerting about the words Robeson speaks near the beginning. Describing the settlement of the USA as prompted by a desire for freedom, he uses the word 'we', part of the film's strategy of including the left and, all too implicitly, blacks in the rhetoric of American destiny. But to hear a black man say, as Robeson does here, 'We crossed the ocean in search of freedom', is to feel an astonishing weight of false consciousness.

Perhaps the most important example of the idea of freedom in Robeson's work is his gradual alteration of the words of *Old Man River,* to bring out and extend its reference to oppression and to alter its meaning from resignation to struggle. The stages in this process are instructive. In his first appearance in **Show Boat** in London in 1928, he changed the words in the second verse from '*Niggers* all work on *de* Mississippi' to '*Darkies* all work on *the* Mississippi', and throughout the song he tended (though with more insistence as his status as a performer increased) to correct the purportedly Southern black English of the lyrics – 'old' for 'ol'', 'and' for 'an'', 'with' for 'wid', 'along' for 'alon', and so on. He did sing, in London, in the New York 1932 production, in the 1935 film and on a mid-thirties British recording, the rest of that verse as written, with words sharper in their reference to black labour than they are sometimes given credit for.

Darkies all work on the Mississippi
Darkies all work while the white folk play
Pullin' those boats from the dawn to sunset
Gettin' no rest till the judgement day.
. . .
Let me go 'way from the Mississippi
Let me go 'way from the white man boss
Show me that stream called the River Jordan
That's the old stream that I long to cross.

The Jordan reference takes us back to the 'eternal suffering' reading of the spirituals, and it is left to the chorus (and Robeson on the record) to sing perhaps the harshest words in the song, harshest in their reference to both white rule and the black man's hopeless lot:

Don't look up, an' don't look down,
You don't dast make de white boss frown,
Bend your knees, an' bow your head
An' pull dat rope until you're dead.

These words were amplified somewhat by the montage sequence in the middle of the number in the film, where the heaviness of Robeson/Joe's labour is shown and where his aching anger is shown in a shot of him with clenched fists raised to the skies. This treatment is interesting because, contradictorily, the character is constructed as, in his wife's words, 'the laziest man that ever lived on this river', and because the publicity for the character in the 1932 stage version and for the film emphasises his smiling affability.

Refusing white words to describe blacks and white imitations of how they thought black people spoke was already an invasion of mainstream white cultural production, but in his concerts and later radio appearances, for which he was always asked to sing *Old Man River,* he changed the words more radically. In the opening verse as originally written the singer wants to be *Old Man River* because the river does not worry about freedom; in Robeson's version, that is precisely why he does not want to be *Old Man River:*

original	*Robeson*
Dere's an ol' man called de Mississippi	There's an old man called the Mississippi
Dat's de ol' man dat I'd like to be	That's the old man I don't like to be
What does he care if de world's got troubles?	What does he care if the world's got trouble?
What does he care if de land ain't free?	What does he care if the land ain't free?

Singing 'I don't like' instead of 'dat I'd like' puts an emphasis on the notes that interrupts the easy run of the phrase, and Robeson if anything emphasises this, giving it a bitterness that suggests a resentment that he has himself been identified with the eternal river of indifference to suffering. Further, the heavy tread of the 'lento con sentimento' original, while it can hardly be avoided altogether, is to some extent broken by taking the whole thing at a brisker pace, which suits the further changes to the words.

In the bridge passage and final statement of the song, the original and Robeson versions show a shift from suffering and resignation to oppression and resistance:

original	*Robeson*
You an' me we sweat an' strain	You and me we sweat and strain
Body all achin' an' racked wid pain	Body all aching and racked with pain
'Tote dat barge!', 'Lift dat bale!'	'Tote that barge!', 'Lift that bale!'
Git a little drunk an' you'll land in jail	You show a little grit and you land in jail
Ah gets weary an' sick of tryin'	But I keep laughin' instead of cryin'
Ah'm tired of livin' an' scared of dyin'	I must keep fightin' until I'm dyin'
But ol' man river He just keeps rollin' alon'.	And old man river He'll just keep rollin' along.

These changes in the words of *Old Man River* are interventions in one of the most popular show tunes of the time; they mark a political black presence in a mainstream (i.e. white) cultural product. This sense of being a black presence in a white space marks every stage of Robeson's career outside the theatre and films – Robeson playing a white man's game (football), Robeson, the only black face in a white fraternity (at Rutgers) and in a white profession (the law), Robeson's white-style home in Connecticut, Robeson posing before the stars and stripes, in (white) American uniform. Most significant of all – but then, like the rest, most ambiguous too – was Robeson with Lawrence Brown in concert, this indelibly white cultural space confidently occupied by two black performers. To many observers, especially subsequently, such moments and images show Robeson as a white

Paul Robeson at Columbia: 'the only blac face in a white professio

man's nigger, sacrificing his specifically black cultural heritage to the codes and conventions of white culture. The (1942) **Look** magazine spread of him and his family at home could easily give offence in these terms. Yet it could also be seen as inspiring to other black viewers and readers, and, partly because of this inspirational quality, as uppity, threatening and offensive to whites.

Paul Robeson singing at a concert in Moscow with Lawrence Brown, 1936

It is not until the late forties that Robeson began to get his come-uppance for his politics. The scale of it is striking even for the era of McCarthy – massive rioting attended his open air concert at Peekskill in 1949, he was subsequently unable to find concert halls in the USA willing to let him perform, and he could not perform outside of the USA between 1950 and 1958 because his passport was taken away. All this lies after, and therefore outside, the period covered in this chapter, but the potentially disturbing aspects of Robeson – his physique, his moral character, his level of achievement, his association with ideas of freedom, his occupation of white cultural spaces – were always present. The way the white media handled him, and especially the cinema, suggests just how disturbing all this could be felt to be.

Body and Soul

Before examining the strategies of the white media for handling Robeson, and as both a prelude to that and a recapitulation of what has been said so far, I want to look in some detail at a film that is differently placed in relation to these problems and yet is profoundly caught up in them. This is **Body and Soul.** It is in a different position from the rest of Robeson's stage and film work partly by virtue of its early date, 1924, when Robeson was only just becoming established as a star, and partly because it was written, directed and produced by the black film-maker Oscar Micheaux. I am not claiming that because a film-maker is black (or a woman, or gay) his or her work will necessarily be more truly expressive of black (or female, or gay) experience, much less of progressive views of that experience. But it is likely to be couched in specifically black (or female, or gay) subcultural discourses, to be in at least a negotiated relationship to mainstream discourses. Not inevitably, but likely; and, as Thomas Cripps (1977, p.193) observes, **Body and Soul** is remarkable because it does bring together 'alternative life-models', specific to black American existence, 'close-packed upon each other in competition'. In other words, Cripps suggests it has an inwardness with the different contradictory ways (images and practices) that black people had of making sense of their situation to a degree no white-made film ever did.

Body and Soul is not a film that is easy to get to see. Cripps suggests that there were two versions, the one that Micheaux originally made, and a revised version made under pressure from the New York censor board, whose objections, according to Cripps, were 'much the same (as) the NAACP would have given' (ibid., p.192), namely, its rather lascivious depiction of ghetto low life. The version preserved at George Eastman House in Rochester, New York, is presumably the latter, although it bears only some resemblance to Cripp's account of the film. This version, and the fact that Micheaux had to adapt his film to white ideas, itself encapsulates the dilemma of constructing such a film. It does not just register, as in Cripps's reading, the alternative life-styles of ghetto society itself but also the problems of how to represent that society, and in particular how to represent black sexual mores.

This is also the problem in Paul Robeson's image at this stage. It is quite clear that he was thought to have 'sex-appeal'. Some found it in his voice.

> *The best description I ever heard of Robeson's voice was from Norman Haire – but unfortunately it is unprintable, since sexual imagery in this country is* **verboten,** *in spite of the fact that sex is life, and all art sexual* (Mannin, 1930, p.158).

Eslanda Goode Robeson's 1930 biography also stresses how much women found her husband sexually attractive. This is never so powerfully expressed in later references, but in addition to the brute/rape motif noted above, the idea of a less aggressive, but still potent, sex appeal was enough to make the Art Alliance

Paul Robeson with Peggy Ashcroft in **Othello** (1930)

of Philadelphia return their commissioned nude statue of Robeson to the artist, Antonio Salemme. It is interesting to note the differences in the costume designs for Robeson's three Othellos, all working within the all-purpose, Tudor-theatrical style. The 1930 version has him in tights, with sensuous tops of fur and flowing velvet or puff-sleeved brocade; legs and inner thighs (if not exactly crotch) are revealed below, chest and arms amplified above. In the 1943 production, on the other hand, he is in flowing robes that wholly conceal the shape of the body, or leave only the arms bare in the murder scene. The same is true of the 1959 Stratford production. There is too a difference in the overall feeling conveyed by him in production photos. In 1930, he is glowering, smouldering; in 1943, troubled; in 1959, anguished, angry. Though it would be wrong to read him in the 1930 photos as just sexual, sexuality is present there to a degree not true of the others.

In the context of Robeson's functioning within white discourses, the gradual elimination of this sexuality from his image can be understood as a further aspect of the need to deactivate, lessen the threat of, his image. **Body and Soul,** however, suggests that there is a tension surrounding this within black discourses too. There is a central dilemma in most black thought of the period, especially that coming out of the Harlem Renaissance – is celebrating black sensuality insisting on an alternative to white culture or, on the contrary, playing into the hands of white culture, where such sensuality could be labelled as a sign of irrational inferiority and more grossly read as genital eroticism, as 'sexuality'? In **Body and Soul** this dilemma is worked through the figure of Robeson, who, in the available archive version anyway, plays two roles, both sexually ambiguous.

One role is Isiah. An opening shot of a newspaper item announces the escape of a prisoner posing as a preacher; there is then a cut to Isiah/Robeson seen from behind. From the start then the film gives the audience a position of knowledge in relation to Isiah – we know he is bad but his parishioners do not. The major narrative of the film concerns Isiah's misuse of his position and the adoration of his congregation to get free liquor and access to Isabelle, the daughter of one of his most loyal admirers, Martha Jane. His rape of Isabelle and her flight in shame to Atlanta, where Martha Jane finds her and learns the truth, lead to Isabelle's denunciation of him in church and the congregation hounding him out of town.

Martha Jane and Isabelle represent the familiar melodrama values of poor-but-honest, toiling and God-fearing folk.

Body and Soul (1924)

This is established in the very first shot of them, the one in bed, the other ironing, which is cross-cut with shots of Isiah talking with the local liquor salesman, threatening to preach against drink if the salesman doesn't keep him supplied with it. This elaborated use of editing is characteristic of the film's method. A later example occurs in the scene between Martha Jane and Isabelle in Atlanta; Isabelle is explaining what kind of man Isiah really is, and this is done through a series of flashbacks. In one, Isabelle has been left by Martha Jane with Isiah, ostensibly so that Isiah can persuade her to give up her wish to marry the man she loves, Sylvester. Isiah attempts first to rape/seduce Isabelle (not for the first time), and then gets her to produce the family Bible (which we already know is where Martha Jane keeps their savings). Three close-ups follow, one of dollars in Isiah's hand, the next of Martha Jane ironing, the third of her picking cotton. This is followed by the title 'Blood Money'. The editing no less than the story is generically classic melodrama, a stark opposition of good and bad, the pure women at the mercy of the heartless villain, the honest poor abused by the greed of the strong.

Yet this is complicated by the fact of Isiah/Robeson's sex appeal. It is not just that Robeson had it, or that, in Cripps's words (ibid., p.192), 'Robeson fairly oozed . . . sexuality', but that we are repeatedly encouraged by the film to feel that sex appeal.

Cripps (ibid.) refers to 'tight close-ups that tilted up to capture a virility long missing from black figures'. In the rape scene – presented as Isabelle's first memory/flashback to her mother in Atlanta – we are placed with Isabelle, and although the purpose of the flashback is to reveal Isiah's dastardly duplicity, it greatly emphasises the sensuality of the experience. Isiah and Isabelle are hiding in a shack after their horse has broken loose from their buggy during a storm. Isiah leaves the room (exiting screen left) so that Isabelle can undress for bed. There follows a sequence of cross-cutting of Isabelle's bare head and shoulders and his feet walking slowly forward (rightwards) in the passage. We have the eroticism of her undressing and nudity, and the tension of his making his way towards her. As we see her in full light, in a classic 'beauty' pose, and see only his active feet, we might seem to have here a standard identification with the male hero in his quest to get the beautiful female object of desire. Certainly we can easily place ourselves thus in relation to the events; but we are invited to place ourselves with Isabelle – it is her flashback – and the final shot, when Isiah enters and before the discreet fade, is a shot of him smiling at her/the camera, *not* of her cowering or responding to him/the camera. Moreover, this shot of him emphasises his flesh by his shirt being opened at the neck (the only time in the film) and by the fact that it is in an iris, a form of framing that often connotes 'the loved one' in silent films (an analogy with the shape of nineteenth-century pocket portrait photographs – compare D.W. Griffith's use of this in relation to Elsie/Lillian Gish in **Birth of a Nation**).

One could not really construe the film as saying that Isabelle wanted to be raped. Her virtue and his evil are quite clear in the narrative context; but the immediate treatment of him in the scene yields to the notion of his sexual attractiveness, even in a rape context. His attractiveness is anyway clearly marked at the beginning of the film. He is the only good-looking male character (bar one significant one) in the film. The first scenes show him with a businessman, who has luridly made-up lips and bright, dirty Jim Dandy clothes, and with a character called Yellow Curley, whose face looks as if it has been plastered with chalk. Both these characters look as corrupt as they are. Beside them, Isiah/Robeson looks the strikingly handsome man he was. He also looks blacker than they do. The businessman's lips suggest an attempt to alter their shape to conform closer to white norms, while Yellow Curley looks a classic stereotype mulatto. Isiah/Robeson is not only blacker in feature, but is not trying to look white either.

We have then in Isiah a character who is unquestionably bad, and yet very attractive because he is so black. It was the lot of black women stars to become known for their beauty only to the degree that they were fair. This is true of several of Robeson's co-stars, including the actress playing Isabelle in **Body and Soul,** as well as Fredi Washington (**Black Boy, The Emperor Jones** (film)), Nina Mae McKinney (**Sanders of the River**), Elizabeth Welch (**Song of Freedom, Big Fella**) – and, most notoriously, Lena Horne. This has not been true of black male stars, although between Robeson in **Body and Soul** and the appearance of stars such as Jim Brown, Richard Roundtree and Billy Dee Williams, there are only two black leading men stars, Harry Belafonte and Sidney Poitier. The question of the sexuality of their image, as of the blackness of it, is a vexed one, but neither were allowed the kind of directly, smoulderingly sexual appeal of Isiah/Robeson or of Brown, Roundtree and Williams.

One further element of **Body and Soul** raises, and confuses, the fact that Isiah/Robeson can be taken as attractive: the only other attractive man in the film, Sylvester, the man Isabelle wishes to marry, is also played by Robeson. Sylvester/Robeson is a much less developed character. We are first introduced to him in an iris shot, walking through woods; this is a cut from a close-up of Isabelle, indicating not what she is looking at but what she is thinking of. (It may even be an important convention that male objects of female desire are introduced this way, thus making the desire more distant and spiritual; compare the introduction of Ahmed/Rudolph Valentino as the object of Yasmin/Vilma Banky's thoughts in **The Son of the Sheik** (see Dyer, 1982a, p.2512).) This shot/thought occurs in a scene at Martha Jane's home, while she and Isiah are talking – but no title indicates who this is (the title may be lost, of course). For all we know, it could be Isabelle thinking of having seen Isiah romantically: the confusion of finding Isiah/Robeson attractive has set in. Sylvester/Robeson only occurs a few times in the rest of the film, chiefly as a brief reminder of who Isabelle really loves. Only at the end does he come into his own. He has invented something (it is not clear what), his invention has been accepted, he has enough money to marry Isabelle – the two of them and Martha Jane are grouped before a portrait of Booker T. Washington, the apostle of just such a black enterprise as Sylvester/Robeson represents. Isiah/Robeson's power was his sex appeal; Sylvester/Robeson's is his respectability.

But there is one further twist. Just before this final scene, there has been a climax to the Isiah story. Fleeing from his congregation, he finds himself in the forest, pursued by a small boy. He turns on him and begins to beat him with a stick. Cut to Martha Jane starting awake in her chair, and the title 'For it was all a dream', at which point Isabelle and Sylvester enter. Isiah then presumably never existed; he was a figment of Martha Jane's dream. Easy enough to read this as psychologically motivated – a mother's fears about the man her daughter loves. But it also represents the wider dilemma the films presents, as did Robeson at that point – the difficulty of stating black sexual appeal. In Martha Jane's dream Sylvester/Robeson's absent sexuality becomes Isiah/Robeson's disturbingly present, attractive, duplicitous, aggressive sexuality.

The rest of Robeson's work, especially in film, works to banish that image even as a spectre in a dream; but it is worth registering how powerfully, disturbingly present it could be. At least a black film-maker, torn apart by the contradictions posed for black culture and politics by the image of an active black male sexuality, registers the problem of handling the image. The later work perhaps handles it all too efficiently, to the point that one might not notice it was there.

Passivity and Pathos

There was no conscious strategy to handle Robeson a certain way. However, even the most progressive white uses of Robeson, such as the avant-garde **Close-Up** group that was behind **Borderline** or the Labour Party-identified group behind **The Proud Valley,** were caught in white discourses that had a way of handling the representation of black people so as to keep those represented in their place. The basic strategy of these discourses might be termed deactivation. Black people's qualities could be praised to the skies, but they must not be shown to be effective qualities active in the world. Even when portrayed at their most vivid and vibrant, they must not be shown to do anything, except perhaps to be destructive in a random sort of way. If narratives, even the residual narratives of songs and photographs, are models of history, then blacks in white narratives may be the colouring of history and the object of history, but not its subject, never what makes it happen.

In this regard, the treatment of blacks bears many resemblances to the patriarchal treatment of women. In addition, the cinematic treatment of Robeson is, with a few exceptions, striking for the degree to which the films deactivate any role that their star may have, in the narrative or even simply in his presence on the screen. As a result there is a contained power about his appearance, a quality that many have found the most moving thing about him.

It is no accident that there are similarities between how black men are represented and how women are depicted. Putting it at its broadest, it is common for oppressed groups to be represented in dominant discourses as non-active. It suits dominance that way, for a variety of reasons – their passivity permits the fantasy of power over them to be exercised, all the more powerful for being a confirmation of actual power; their passivity justifies their subordination ideologically (they don't do anything to improve their lot); their activity would imply challenge to their situation, to the dominant it would imply change.

There is a more specific linkage in US representations of all black people and all women. Many of the nineteenth-century arguments for the abolition of slavery saw blacks and women as sharing a similar nature. At its least attractive, this was a kind of 'you shouldn't kick the weak, you shouldn't mock those inferior to you' argument; but it also came couched in feminist terms (just as there were close connections in political practice between abolitionism and feminism in the mid-nineteenth century in the USA). George Frederickson (1972) refers to this in relation to what he calls Northern 'Romantic Racialism'. Harriet Beecher Stowe is exemplary here, but Frederickson also refers to male abolitionists and pro-feminists of the period. Theodore Tilton, for example, the editor of the **New York Independent,** stated in a speech at the Cooper Institute, New York, in 1863:

> *In all the intellectual activities which take their strange quickening from the moral faculties – which we call instincts, intuitions – the negro is superior to the white man – equal to the white woman. It is sometimes said . . . that the negro race is the feminine race of the world. This is not only because of his social and affectionate nature, but because he possesses that strange moral, instinctive insight that belongs more to women than to men* (quoted by Frederickson, 1972, pp.114–15).

While statements as ideologically explicit as this would not have received universal assent, it would be difficult to overestimate the impact of Stowe's work, and thus ideas like this, on the popular imagination and on the habitual way of thinking about black people. In any event, the treatment of black men, when they are not brutes, constantly puts them into 'feminine' positions, that is, places them structurally in the text in the same positions as women typically occupy.

To argue this through in relation to Robeson, I'd like to look first at aspects of how he is photographed, in pin-ups and photographs, to turn next to the question of his performance style and finally to look at the way he is used in narratives, his function in the overall structure of plays and films and the way he is edited in to the local structure of specific film scenes.

Pin-ups of white men are awkward things. As I've argued elsewhere (Dyer, 1982b), they exemplify a set of dichotomies – they are pictures to be looked at, but it is not the male role to be looked at; they are passive objects of gaze, but men are supposed to be the active subjects of gaze, and so on. Many male pin-ups counteract the passive, objectifying tendency by having the model tauten his body, glare at or away from the viewer, and look as if he is caught in action or movement. This portrait of

Sanders of the River (1935)

King Solomon's Mines (1937)

Robeson taken for **Sanders of the River** is like that. It is particularly instructive to compare his body here with the more snapshot-like photo of him with Kenyatta discussed above – the slack body in the latter is here tautened for the purposes of a classic male pin-up.

However, such treatment of Robeson is untypical. Far more characteristic is this portrait taken for **King Solomon's Mines.** It is in some ways similar to a later portrait of Johnny Weissmuller as Tarzan. Dappled light and vague fern-like tall grass backdrops suggest a deep-in-the-forest/jungle setting. Both create a soft, 'beautiful' feeling, but whereas Weissmuller is posed with his body turning, resting on his arms, Robeson is posed sitting, his arms resting on his knees. Thus Weissmuller seems to be caught in action and his body is tensed, Robeson is still, his body relaxed. Moreover, Weissmuller looks up, in a characteristic pose of masculine striving; Robeson just looks ahead, with a slight, enigmatic smile on his face.

The Robeson portrait has at least as much in common with the early portraits of Marlene Dietrich. Particularly notable is the use of textured clothing that catches and diffuses the light and so blurs (rather than outlining; cf. the discussion in the

Pin-up of Johnny Weissmuller (photo Cecil Beaton, courtesy of Sotheby's London)

Marlene Dietrich

Monroe chapter above) the shape of the body. Like Dietrich, there is perhaps a trace of irony, a sense of detachment. Unlike Monroe, Robeson does not give himself to us, any more than Dietrich does. Like Dietrich though, he is set in place by the image – any movement would spoil the beauty of the composition, the effect of the light. He is pinned in place to be pinned-up – and he has laid his spear and shield aside.

As much like Dietrich as like Weissmuller, Robeson is put into a position of 'feminine' subordination within the dynamics of looking that are built into any portrait-image of a human being in this culture. (These dynamics are more complex than a 'men look: women are looked at' opposition, though that is a fair starting point.) **The Emperor Jones,** the film in which Robeson has the most dynamic role in the narrative of any of his sound films, is also – and perhaps not coincidentally – the film that, at the level of visual treatment, most suggests this kind of 'feminine' positioning. Particularly striking is the constant use of mirrors, to suggest a narcissistic involvement with self that is typically male but always coded as feminine. The first shot of Robeson in the film shows him preening in front of a mirror, followed immediately by a shot of a woman looking at him, saying, 'Um – um, yo sho' is wonderful in them clothes', with a cut back to him looking at himself in the mirror some more. By the time he is ruler of the island, he has installed mirrors all over his palace. The introduction of him in full Emperor regalia is done with a shot that tracks down from the top of an ornate mirror in front of which Jones/Robeson is dressing himself; when he goes to see his court he walks along a hall with mirrors on the wall, stopping for a moment in front of each one to catch his reflection. This is part of the film's mockery of Jones; male concern with display, finery, appearance, looking good, is often mocked through this kind of mirror imagery and its connotations of narcissism, and the thrust of the jibe is that such concerns are 'feminine'.

The analogy of this visual treatment of Robeson with that of women might also suggest that these images, like pin-ups of women, are heavily coded in terms of eroticism. Undoubtedly sex appeal is, powerfully, one of the ways that such images of Robeson can be taken; but much pin-up and portraiture of him works out of a tradition that is more ambivalent on this question, that of the idea of the classical nude.

Robeson is often taken, in photographs and in one-shots in films, in poses that derive formally from the statuary of classical antiquity. This 1926 photograph by Nicholas Murray has many of the formal features of the work of, say, Praxiteles or a nineteenth-

Nude study of Paul Robeson by Nicholas Murray

century neo-classical sculptor such as Canova, or many between and since – the stance with the weight on one slightly bent leg, the other held straight; the head held downwards, sideways; one hand turned inwards, the other outwards, both held halfway to a grasp; one shoulder higher than the other; a sense of flowing lines through the body.

The significance, in our period, of this formal similarity to the art of classical antiquity in the treatment of the nude is that it makes possible an appeal to the idea of timeless, immemorial ideals enshrined in the classics. This appeal can then be used to defend the practice of representing the naked human body in a period when such representation is widely deemed lewd or immoral. In order to be able to go on producing images of naked people at all, art practice had to produce a discourse that denied the erotic dimension of such images, that insisted classical style nudes transcended sexuality and were a celebration of the ideal beauty of the human form, 'proportion, symmetry, elasticity and aplomb' in Kenneth Clark's words (1960, p.30). This defence of the nude effects a double articulation – it is both a way of producing potentially erotic images while denying that that is what is being done, and also a way of constructing a mode of looking at the naked form (as generations of art school students have) 'dispassionately', without arousal.

The classicism of still images of Robeson fitted with a quite general perception of him as cast within 'the heroic mould'. Elizabeth Sergeant referred to both physical and ethical dimensions. Physically, Robeson was 'of noble physical strength and beauty . . . like a bronze of ancient mold' (1926, p.40), while ethically 'unlike most complex moderns, [he] does not appear half a dozen men in a torn and striving body . . . he is one and clear-cut in the Greek or primordial sense' (1927, p.194). Similarly, Edwin Bancroft Henderson (1939, pp.33–5) enthused over another black male with whom Robeson was linked, Joe Louis, in a disconcerting mixture of racial praise:

Joe has the physique of a Greek god. His color appears to be a golden bronze. Few Negro fighters have shaded lighter than he.

The pure beauty ideology was clearly expressed by Antonio Salemme, who did the nude statue of Robeson that upset the moral-artistic guardians of Philadelphia. His words to Robeson, reported in Eslanda Robeson's biography and repeated in Shirley Graham's, are pure art school ideology.

> *All that we are exists in the body – mind, spirit and soul. The human form takes its beauty from all these. Its lines may suggest full, glowing life or dormant, empty life; the muscles may suggest fine, free, powerful movement, or calm stillness and peace. The body has harmony, rhythm and infinite meaning* (Robeson, 1930, pp.81–2).

In this rhetoric, art is sex blind, the human body ungendered; but in practice it is the nude female whose muscles suggest 'calm stillness', the nude male's that suggest 'fine, free, powerful movement' – the white nude male, that is.

In the classical period proper it was possible to represent the athletic male nude body as an object of contemplation, but by the nineteenth century (and long before), if the naked male body was to be represented at all, it had to be doing something, in action. Westmacott's **Achilles** or Eakins's studies of boxers were not just images of bodies capable of actions, but actually in action. Even George Washington's statesmanship translates, in Greenough's statue, into physical action. Yet Robeson's still images, though we see an athletic, muscular body, tend to show a potential for action only, not action itself. In Murray's standing portrait, the head is bowed more than in the classical model,

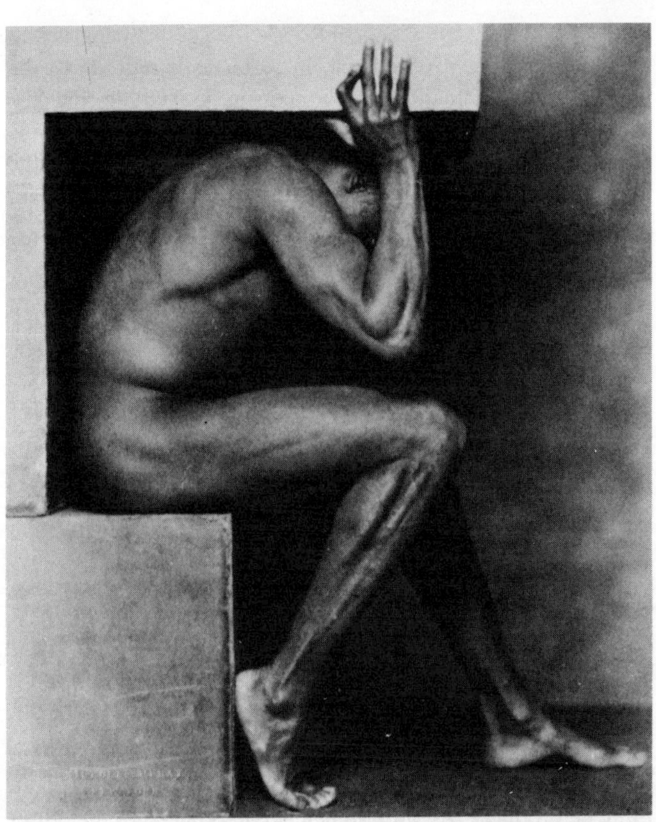

Another nude study of Paul Robeson by Nicholas Murray

perhaps suggesting the weight-of-sorrow view of the black man's lot. More striking is another study by Murray of Robeson with his hands holding on to a slab (of concrete?). The position of his right arm and leg, nearest to us, show their muscularity – but the pose does not actually suggest the model is doing anything. The bent leg is not giving leverage, the arms are not holding up the slab but merely holding on to it. The body's potential for strength is revealed, but if anything the use of that strength is denied by the positioning of the legs and arms. Not only are they not doing anything, there is nothing they could do in that position – the concrete(?) blocks are fixed. What the blocks do is confine the model, he is bowed over and contained by them. Even more is the sorrowful weight of oppression suggested, but not the heroic resistance or overthrow of it.

Finally in the Murray studies, not only do we not see the penis (moral censorship? fear of showing a black man with phallus?); we also don't see the face. That part of the body that carries the greatest burden of meaning, in terms of personality and expression, is hidden.

The visual treatment of Robeson suggests analogies with the visual treatment of women, in so far as it reproduces the feeling of subordination of the person looked at; but the 'classicism' of the approach plays down and may even dispel altogether the eroticism of the images. What is produced is the idea of passive beauty. This corresponds also to aspects of performance:

There is no remark that is so disparaging to the Negro actor, singer, musician, as the one – often intended as a high compliment – that he is a natural born actor (Edith Isaacs, **The Negro in the American Theatre).**

It is interesting to look again at Salemme's quoted words to Robeson. When the latter said that he did not think he could 'pose for a sculptor. I don't know how', Salemme replied,

Posing is not what I want. Your body is something beautiful to behold. It is expressive; it has intelligence. I'd like to work it in bronze.

Edwin R. Embree's version of these words is, 'Good God . . . you don't need to pose; just take your clothes off and stand there' (1945, p.251). Robeson's beauty, even his expressivity and intelligence, become structured as natural possessions, not something he produces. As with women, so black men are the object of, in Mary Ellman's words, 'the celebration of thoughtless achievement'. This sense of Robeson just 'being' beautiful, wonderful, expressive underlies much of the rapturous critical reception to him as a performer.

In addition to remarking on his size and magnificent physique, critics and observers sing the same refrain:

> *Robeson . . . is one of the most thoroughly eloquent, impressive and convincing actors that I have looked at and listened to in at least twenty years of professional theatregoing. He gains his effects with means that not only seem natural, but that* **are** *natural. He does things beautifully, with his voice, his features, his hands, his whole somewhat ungainly body, yet I doubt that he knows how he does them* (George Jean Nathan, review of **All God's Chillun Got Wings, American Mercury** July, 1924; quoted by Noble, 1948, p.141).

In the same article Nathan risked the generalisation that 'the Negro is a born actor, where the white man achieves acting'.

Similarly,

> *Paul Robeson is a natural artist, completely lacking in self-consciousness, or affectation* (Mannin, 1930, p.158).

> *Robeson possesses a beautiful, natural, unforced baritone voice of great volume . . . Ivor Brown in* **The Observer** . . . [wrote] *'Mr Robeson's Ebon Othello is as sturdy as an oak, deep-rooted in its elemental passion . . . One thinks of a tree because the greatness is of nature, not of art'. (Cuney-Hare, 1936, pp.372–3).*

Or compare black writer, Edwin R. Embree, in his book celebrating black people who have made it against the grain of American society, **Thirteen Against the Odds:**

> *He has travelled only two of Shakespeare's roads to greatness. He was born great, and – almost against his easygoing inclination – he has had greatness thrust upon him* (Embree, 1945, pp.249–50).

Robeson himself sometimes gave credence to this view by his reported descriptions of his own approach:

> *I'm not a great actor like José Ferrer . . . All I do is feel the part. I make myself believe I am Othello, and I act as he would act* (Reported by Jerome Beatty, 'America's No. 1 Negro', **The American Magazine,** vol. CXXXVII, no. 5, May 1944; quoted by Schlosser, 1970, p.182).

This way of reading Robeson's performances as unproduced fitted with the fact that his acting style used a minimum of performance signs, or rather used stillness, a small number of gestures, intensification of the voice rather than more movement, elaborated gestures and vocal gymnastics. (I have already discussed something of this aspect in relation to his singing style and its purity and simplicity.) Stillness, a minimum style of performance – these can be treated as deliberate choices, to do with a refusal of theatrics or else a method for producing intensity, concentration. Robeson could certainly be read in the latter way –

> *It was Mr Robeson's gift to make [the spirituals] tell in every line, and that not by any outward stress, but by an overwhelming inward conviction* (**New York Times,** 20.4.25, p.2)

but he was more generally read emblematically, his stillness a passive, inactive embodiment of value.

In the 1942 **Othello** he could be seen as the victimised symbol of nobility, simplicity, dignity, partly through the contrast between José Ferrer's Iago, 'all movement and all over the stage

1928 London production of **Show Boat:** *Can't Help Lovin' Dat Man*

at once' and his Othello's 'sculptured stillness' (Hamilton, 1974, p.108). That this is also what some critics expected Robeson to be (when they were not hoping for aboriginal passion) is revealed in Michael Mac Liammoír's comments on the 1959 production, where even the few gestures Robeson did use were felt to interfere with his emblematic stasis. Mac Liammoír suggests that it would have been better if Robeson was just allowed to be

> *immobile and magnificent as the Sphinx . . . allowing him to rely for interpretation on the profound reality of his voice* (**The Observer,** 12.4.59).

Robeson's immobile performance style is suggested by this production still from the 1928 London production of **Show Boat.** It is the *Can't Help Lovin' Dat Man* number, in which Magnolia, the daughter of the white owner of the show boat, is singing the 'nigger' song that the mulatto Julie has taught her. Magnolia's movements are a conscious imitation of 'nigger' dancing; everyone else, bar Robeson/Joe and Alberta Hunter/Julie, are performing with minstrelsy derived, cavorting or bending movements. Hunter/Julie, not yet revealed as a mulatto, remains somewhat more ladylike (= white). Robeson/Joe stands in a rather classical pose, the emblematic onlooker. He refuses the coon style performance, and is left with the purity and simplicity of mere presence.

This is also how he is used throughout both play and

1928 London production of **Show Boat:** Finale, Act One

film. The role is sometimes described as being akin to a Greek chorus, commenting on but not participating in the action. This may be structurally accurate, but the Greek chorus is anonymous, characterless, representing the universal everyperson of implied author and audience; Joe/Robeson, on the other hand, is, as character, a working black man in the racially segregated world of the South, and, as performer, appearing in an expensive show before predominantly white audiences. Though *Old Man River* appeals to the universal, it roots the experience specifically in black toil and tribulation. Moreover, Joe's emblematic presence is most called upon at specifically racially sensitive moments in the

1928 London production of **Show Boat:** the miscegenation scene

play and film. He is just there throughout the first half, but his presence is emphasised when the plot hinges on racial questions, especially the discovery that Julie is a mulatto and her banishment from the show boat. Compare his positioning in the finale to Act One (1928 production), where he stands out by virtue of the simplicity of his clothing but not because of where he is standing, with his positioning in the miscegenation scene, a quite startling placing that draws attention to his inactive looking on at this racially tense moment.

In the film this is emphasised by the use of editing. When Ellie comes rushing into the rehearsal, full of her secret (that Julie is not 100 per cent white), there is a cut in of a close-up of Joe/Robeson, looking and listening; when the sheriff enters to arrest Julie, another cut-in of him; when Steve, who has previously cut Julie's hand and sucked some blood from it, declares that he too has Negro blood in him, another cut to Joe/Robeson, who with one movement of his eyes and head conveys a wealth of unfathomable meaning; finally, when the sheriff declares that folks in the area would be put out to know there are white people acting with black, the film cuts again to Joe/Robeson's face. Quite precisely, the film, even more it would seem than the show, stresses Joe/Robeson as an emblem of racial suffering. Both staging and editing register awareness of racism in Robeson's passive observation of it.

Joe/Robeson only once does anything in the plot, and that is to go through the storm to fetch the doctor for Magnolia's confinement. This action is in the film only, and shows courage, but entirely in the service of white destinies. After the early plays and films, Robeson seldom had a role that involved actions that had an effect on the plot, unless it be in white interests. The role of leader of his people – taking them on a trek to find salt in **Jericho,** organising them into a farming collective in **Tales of Manhattan** – might give him such a narrative function, but even as a leader, he is effectively ineffective. In **Sanders of the River,** apart from quelling King Nofolaba's warriors early on as the British administration requires, being king involves little but sitting about in regal regalia. In **Basalik,** as a native chief, critics complained that he had 'little to do but stand about and look noble' ('IB', **Manchester Guardian,** 9.4.35; quoted by Schlosser 1970, p.156). When not playing a leader, his part becomes little more than that of helper to whites in their problems, whether domestic – the runaway boy in **Big Fella,** who, when he is returned to home, sickens until Banjo/Robeson comes to his bedside – or domestic and industrial, as in **The Proud Valley.**

This film displays a white community's easy acceptance of a black worker, yet unconsciously demonstrates the terms of that acceptance. As far as work goes, David/Robeson can be one of the men but neither a leader – he refuses to go in with the other men to see the Minister responsible for mining – nor a survivor – when a group of the men are trapped underground it is he who lights the dynamite, knowing it will free the others and kill him. In both cases, David/Robeson makes the decision – at the ministry he tells the others to go on in without him, and before their feeble protestations, says, unanswerably, 'Oh no, I wouldn't be much use to you'; down the mine, lots are drawn as to who is to light the dynamite fuse and Emlyn does so, but rather than let him do it, David/Robeson knocks him out and lights the fatal fuse himself. He participates in his own subordination and sacrifice. It is uncomfortably close to what Elizabeth Sergeant (1926, p.41) called 'that self-denying, passive, deeply impressionable Negro essence' that makes black people such good performers.

In domestic terms, David/Robeson's role in **The Proud Valley** is again purely as helper, sorting out the problems of the whites around him, but having none of his own. In one sequence, he acts, on his own initiative, as go-between for Emlyn and his girlfriend Gwen who have fallen out; the model for this, including the light comedy of the subterfuge and manipulation involved (e.g. telling Gwen it is all over with Emlyn so that she will go to him fierily and tell him that it is not) is classical comedy, Robeson playing the servant role to the silly lovers. Despite being the star of the film, there is no question of him being a lover.

Not only in overall plot terms, but also in the construction of individual scenes, David/Robeson is marginalised. In a sequence quite early in the film, David/Robeson has just been introduced to Dick's family – there are questions of whether there is room for David to be a lodger, of Emlyn and Gwen's courtship, of the forthcoming Eisteddfod. The scene begins with an establishing shot, with David/Robeson sitting to the right of the table and the others ranged round it. However, once discussion starts, we move not to close-ups of speakers but to two- and three-shots with David/Robeson excluded, so that the group becomes defined as the whites. Only at the end of the sequence do we have the establishing shot repeated. David/Robeson's marginalisation is particularly noticeable at emotional moments in the film – he is the star and ostensibly the main character, yet when the bell announces the pit disaster, there is a quick series of close ups, of Gwen ('The pit!') Emlyn ('Dad!') and Mrs Parry ('Oh, my God!'), which put all the emotional drama on to Dick, even though

David/Robeson is also down the pit at the time. When the miners set off on their march to London, each has a loved one to kiss him good-bye save for David/Robeson who has to hang about awkwardly in the background.

The most sustained use of Robeson in such a way is seen in **Borderline.** Here the emblematic approach meshes with the theoretical position that informs the film, derived from Soviet theories of montage. **Borderline's** use of Robeson might even be the same as Sergei Eisenstein's would have been, had they made a film together as they planned. Vladimir Nizhny's description of how Eisenstein spoke in his classes of using Robeson suggests an essentially emblematic function.* He showed the class photographs of Robeson, and referred to his 'rich temperament . . . physique and marvellous face' (1962, p.27), and then talked about a scene in which Dessalines storms a castle. In discussing how to achieve maximum impact, Robeson, the performer, becomes a plastic element, important for his 'temperament' and emblematic blackness –

> *when . . . a candelabrum, with lighted candles to boot, sparks blazing and flickering from its pendants, is raised by a man of gigantic stature, with dark face and flashing eyes and teeth (remember Paul Robeson), this will be not only effective, but a veritable climax to Dessalines' indignation* (ibid., p.58).

The ideological-aesthetic justifications for this emblematic use of performers are well known – the desire to make crowds not individuals the hero of the (hi)story, the use of individual performers as types representative of social groups. In the context of **Black Majesty,** where all performance might have been approached in this way, this might not have made Robeson's appearance in the film different or passive (– though Eisenstein's words do suggest the frisson of the white contemplation of the huge black man). In **Borderline,** the same approach is more problematic.

Borderline was written and directed by Kenneth Macpherson, a member of the avant-garde group associated with the film journal **Close-Up** which had links with the Harlem Renaissance (producing a special issue on blacks and film), the literary and quasi-feminist avant-garde (H.D., Marianne Moore, Gertrude Stein, Dorothy Richardson), psychoanalysis (Barbara Low, Mary Chadwick, Hans Sachs and Freud himself), as well as formalist film theory (publishing the first translations of Eisenstein among others). (For a full account of this see Friedberg, 1980–1).

*There is some dispute over what part Robeson would have played. The film was to be called **Black Majesty** and to deal with the Haitian revolution in the nineteenth century. Nizhny says Eisenstein wanted Robeson to play Dessalines, but in a footnote to the text Herbert Marshall says that Robeson thought Eisenstein wanted him to play Toussaint l'Ouverture, while he, Marshall, is certain it was Henry Christophe. In any event, what is interesting is Nizhny's account of how Eisenstein intended to use Robeson. Even if Nizhny reports this inaccurately, this is no great matter here; it is not a question of what Eisenstein did or did not think, but rather of the way Robeson figures in the discourse of Soviet montage theory.

Although constructed more insistently through the chains of association and rhythmic effects of its montage, **Borderline** does have a plot, concerning a white couple, Thorne and Astrid, and a black couple, Adah and Pete (Eslanda and Paul Robeson), all of whom live in the same inn in a small village. There is an implication that Thorne and Adah have been sleeping together. Astrid and Thorne fight, over his infidelity perhaps but also out of the jadedness of their relationship. Thorne stabs Astrid. The racial prejudice of the other villagers mounts; Adah goes away, leaving Pete a note saying it is all her fault. Pete is ordered out of the town by the mayor, but before he goes, Thorne, who has been acquitted of murdering Astrid, comes to the station to say good-bye. This bare outline of a deliberately elliptical plot already suggests how little an active role the Paul Robeson character has in the narrative. The highly complex use of montage only reinforces this inactivity.

The film is organised around the basic antinomy of black and white at every level, aesthetic, metaphorical, ethical, ethnic. The photography exploits black and white stock for dynamic visual effects, effects, that is, based on an aesthetic of contrast or clash between dark and light areas and shapes. Equally the film sees black and white people antithetically. The traditional Western moral dichotomy of black and white is also maintained,

Borderline (1930)

but reversed in its conflation with race – that is, in **Borderline** racially black equals morally white and vice versa. These antinomies are further expressed in other elements of film style – black characters are shot still and in repose, in visually simple compositions; white characters are shot in frenzied movement and gesticulation, in complex compositions. As a result, black performers (the Robesons) do not do anything, whereas white performers, especially H.D. as Astrid, are constantly, albeit destructively, active. All this recalls the white alienation versus blackfolks spirituality opposition that runs through discourses on blackness in the twenties and thirties.

A good example of **Borderline's** method occurs early in the film. Thorne and Astrid are in their room, bored and tetchy; Pete is in his, just there. The cuts in the Thorne-Astrid scene focus on jagged, quick movements, Astrid clutching at her fluttering tea-gown, Thorne playing with a knife. The cuts in the Pete scene are a series of close-ups of Pete/Robeson's body in repose – his profile, his hands, his body stretched out on the bed. Neither scene is primarily concerned with narrative, rather they contrast the feeling of the black and white temperaments.

Shots of Paul Robeson throughout further emphasise his passive, emblematic beauty – his dark, smiling face contrasting with bulbous white clouds; shots of him motionless on a hillside intercut very fast with a tumbling waterfall (this, again, being also a cutting between a predominantly black and still and predominantly white and fast image); a close-up of him laughing, with a white flower in his ear, and so on.

In addition to the racial and aesthetic significance of this photography, there is an implicit sexual dimension. Although hardly an explicitly gay film, the construction of the narrative and the emphasis on Robeson/Pete's passive beauty both suggest that he functions as an object of desire in the film. There are touches of gay subcultural iconography – the dyke style of the innkeeper and her woman friend, for instance, and the piano player with the photo of Pete/Robeson on his piano. More importantly, the last scene in the film with any emotional weight is that between Thorne and Pete, looking at each other and, in extreme close-up, shaking hands. At one level this suggests friendship transcending race, but it is also a friendship transcending adultery and death. After it we have a shot of the flower Pete/Robeson was wearing in the bar, dead in its glass, and a shot of Thorne sitting alone beneath a cherry tree on the mountain. The flower imagery evokes love and the death of love between the two men.

Borderline's use of Robeson can be seen as a product of

aesthetic theory, blackfolks ideology and gay sensibility. It is different from the use made of him in other films, but it is still broadly within the general discourse that deactivates the black person even while lauding her or him.

The only exception to this kind of treatment is Robeson's roles in left theatre. As Toussaint l'Ouverture (1935) and John Henry (1940), in the plays of C.L.R. James and Robert Roark respectively, he played the traditional active hero of a narrative. Two socialist plays, **Stevedore** and **Plant in the Sun,** developed a different kind of protagonist-narrative relationship. Neither play was written with Robeson in mind, and he appeared not in their original US productions but in those by the Unity Theatre in Britain. Both are concerned with racial questions, but their emphasis is on the need to sink racial differences in the recognition of class identity, neither black nor white but workers.

In each case Robeson plays a worker who is both an organiser of his fellows and a victim of discrimination because of it. In **Stevedore** (London, 1935), a wrongful accusation of rape against the Robeson character, Lonnie, is used as an excuse for attempting to lynch him and so curb his union activities; in **Plant in the Sun** (London, 1938), his character, Peewee, is fired for such activities. In both cases his treatment is what galvanises his fellow workers, white as well as black, into action in support of him. **Plant in the Sun,** a lighter play with much humour in it, ends with the action of Peewee's immediate mates, a sit-down strike, inspiring the whole works to join in, thus promoting the tactic of the sit-down strike that had been so successfully used at General Motors in 1936 (see Goldstein, 1974, pp.184–5). Accounts of the play suggest that Robeson/Peewee gives the initial impetus for what happens by articulating the politics of action and by being the occasion of it; thereafter his role is as one of the group. In the London production, this was echoed in the way that he was not given star billing but simply treated as one of the company – the mode of production of the play embodied the ideal represented by the play.

Stevedore, a more sombre piece, ends with Robeson/Lonnie shot by the leader of the white lynch mob, an action which leads first to the blacks turning on the lynch mob rather than taking the quietist stance that many have been urging up to that point, and then by Robeson/Lonnie's fellow white workers joining in and routing the lynch mob. In Karen Malpede Taylor's (1972, p.77) description: 'The curtain falls on a tableau like the Pietà: Ruby holds Lonnie's lifeless body in her lap'. As with **Plant in the Sun,** the Robeson character both articulates the socialist and racial consciousness of the play and is the occasion for realising that consciousness in action. Taylor argues that although each of **Stevedore's** three acts ends 'with a moment of pathos', Lonnie dragged away to be lynched, another character shot, Lonnie murdered, 'the crux of each act, the moment of conflict which determines in which direction the future action will go, occurs as the black community confronts its white oppressors . . . The defeat and deaths of individuals provide three emotional endings. Yet this . . . domestic tragedy is undercut by an epic movement which . . . [represents] a nation struggling to be born' (ibid., pp.77–8). The Robeson character is active in the narrative, but ultimately this action is taken up into the broader sweep of collective action. Whereas the mainstream plays and films render the Robeson character passive and ineffective, the left-wing plays show his individual action as part of collective action in history. Yet it is interesting that **Stevedore** should end on a Pietà – the resonance of the pathos of the kind of black male image that Robeson was elsewhere taken to embody in his singing and person is still drawn on here, perhaps only as a universalising touch, but maybe as a lingering register of the 'romantic realism' of the left.

The image of the heroic Robeson/Lonnie dead at the end of **Stevedore** may also have had a peculiarly powerful pathos because it reworks a certain kind of feeling that is often proposed in Robeson's work. This is the contrast between his terrific potential for action and the fact that it is either curbed or not used; it is the pathos of strength checked, of power withheld, of the beast caged.

Robeson's given physical power was always evident. His sheer size is emphasised time and again, as is the strength presumed to go with it. **Song of Freedom** early on has him picking up a crate single-handed that his white workmates can't manage; when he starts singing, someone says, 'Where did you hear that?' He replies, 'It's been at the back of my head ever since I was a little fella', to which comes the retort, 'When was you ever a little fella?' His size is evoked in the title of **Big Fella,** and used for wry visual humour in the scene where he first meets the tiny Phyllis in **The Proud Valley.** The crate lifting in **Song of Freedom** is like a strong man's turn, as are shifting the boulder in **King Solomon's Mines** or the fallen timbers in **The Proud Valley.** Size and strength were complemented by the power of his voice, its deep, resonant quality. The critic of the **Hartford Daily Times'** description of Robeson's suitability for his role in **Black Boy** sums up the sense of power of Robeson's physical make-up, though the final reference to 'mobility of expression' is interesting as a reference to something more often, as I've argued, curbed in most productions:

> *Mr Robeson's equipment for this role is well-nigh perfect; a physical giant with a voice so deep and rich and powerful, both in speech and song, that it would make a dramatic appeal even without the aid of a mask of extraordinary mobility of expression* (ibid., 12.9.26; quoted by Schlosser, 1970, p.92).

Yet this power was often contained, by being turned into the spectacle of passive beauty, by becoming a turn, by the editing techniques I've discussed, or by the contrast between it and the subject matter Robeson was performing. Here was this big man singing songs that were taken to express sorrow, resignation, humility. Here was this giant playing men humiliated – by their own superstition (**The Emperor Jones**), by their love for a worthless white woman (**All God's Chillun Got Wings**), by schoolmasterly white superiority (**Sanders of the River**). Above all, there was the contrast between the potential bodily power and his actual stillness and, most movingly, between the potential vocal power and his soft, gentle, careful actual delivery of speech and song.

This is very clear in Robeson's earliest recordings, made in 1925 and 1926 and reissued in 1972 by RCA as **Songs of My People.** The voice here, as later, is capable of great power; some phrases and passages are delivered with a full-throated energy; there are already the deep, resonant, sustained notes so characteristic of the sound of Robeson in popular memory. These heavy, forceful qualities are lightened here by a number of factors: a wider vocal range, so that he can move into the tenor register easily and soaringly, without the (exciting) sense of effort he required later; greater use of grace notes, making the overall effect more elegant but also less direct; the choice of as many fast and exultant numbers as slow and mournful ones, sung with an astonishing brilliance, rapid phrases tripping off the tongue in a manner not generally associated with bass singing. At the end of *Get On Board, Little Children,* he sings the refrain once through straight and strong, but then softens his voice (and Lawrence Brown self-pedals the piano) so that the final statement of the refrain is not loud and firm, but muted and delicate. For the last phrase 'many a more', his voice takes on the plummy, rounded sound of parlour balladry on 'many a' and then just fades gracefully on the long-held final note on 'more'. The kind of rousing, affirmative feeling that is usual with this song (and in some of Robeson's later performances of it) is modified by a gentleness and quietness which could also be called damping and restrained.

Many listeners sought phrases that would capture something of this remarkable quality:

> *G.B. Stern said to me that Rebecca West had described Robeson's voice as 'black velvet', but that Van Druten defined the shade as 'mulberry', which did I think the better adjective?* (Mannin, 1930, p.157).

> *He combined with a rich and mellow voice a dramatic restraint and power that seemed to* **hold** *unheard* **thunder** *behind each song* (unnamed article quoted by Ovington, 1927, p.213; my emphasis).

> *Some tones were so deep that they suggested the elemental sound of thunder; others were strangely clear, high, sweet and gentle* (Seton, 1958, p.50).

Cicely Hamilton's description of Robeson in the London production of **The Emperor Jones** articulates the kind of affect produced by this contained, restrained, withheld power:

> *Something of Mr Robeson's success is due, no doubt, to his personality; to his voice, which is soft as well as resonant, to his racial intonation and his size. Above all to his size; there was pathos almost unbearable in the humbling of so mighty a man (***Time and Tide,** vol.6, no.39, September 1925, pp.938–9; quoted by Schlosser, 1970, p.79).

Similarly, Marie Seton's (1958) description of his singing *Sometimes I Feel Like a Motherless Child* in concert:

> *There was something almost painful about this massive man with strong, forceful features speaking in song with such infinitely tender and sorrowful yearning.*

It makes a difference what the source of humbling is – whether contained by the compositional or narrative structures of the text he appears in or held back by his own performance technique – but the fact of it, the moving affect of it, may be the emotional heart of Robeson's cross-over appeal.

At the end of the film version of *Old Man River* (which

Robeson recorded close to the microphone, hence softly, not boomingly), there is an extreme close-up of Joe/Robeson. He finishes singing, and then looks ahead for a few moments – neutrally? sadly? bitterly? – before gradually forcing a smile through his face, so that the shot and number can fade on a sambo grin. It is the kind of moment John Ellis (1982) refers to as something we search out in star performances, that meaningful flicker of expression that we think we see when we have mentally cleared away all the hype and production.

It looks like Robeson the performer having to make Joe the character smile, being forced to play Sambo. It is 'unbearably moving' because it is the humbling of a great singer in the service of a demeaning stereotype. Obviously, what I am offering here is my own reaction as evidence of a possible reaction, but it is of a piece with Hamilton and Seton, quoted above, and many others. It is a response that needs a critique, for it puts me in the position of agonising exquisitely over the fate of a black man, getting off emotionally on the humiliation of a people. That is the kind of price being a cross-over star may involve with some audiences. In any event, when Robeson no longer played the part of power withheld and became more vigorous and harsh in his vocal delivery as well as his opinions, he ceased to be a cross-over star.

Conclusion

> *We, over a period of time, have apparently decided that within American life we have one great repository where we're going to focus and imagine sensuality and exaggerated sensuality, all very removed and earthy things – and this great image is the American negro* (Lorraine Hansberry, **Variety,** 27.5.59, p.16).

> *It did not occur to us that there would be any objection to showing a nude figure . . . The executive committee, however, expressed their apprehension of the consequences of exhibiting such a figure in a public square, especially the figure of a Negro as the coloured problem seems to be unusually great in Philadelphia* (Walter Hancock on behalf of the Art Alliance of Philadelphia; quoted in **Opportunity,** June 1930, p.168).

It is no accident that so much of the argument surrounding Robeson as a star should involve discussion of perceptions of his body and of how his body was photographed, directed, used.

All performers use their bodies and have their bodies used, but in the case of Robeson in the twenties and thirties he was often little more than a body and a voice. In this regard, his treatment is typical of the treatment of black people, female and male, in Western culture.

Representations of black people are one of the primary sites where the problem of the body is worked through. This is not just in the kind of celebration of sensuality promoted by whites and blacks associated with the Harlem Renaissance; it is also central to many of the characteristic white narratives centred on black characters – the rise to fame of the black man through the use of his body in sports; the hysterical treatment of the mulatto, the product of a mingling of blood who lives out in her/his body the racial confusions of a society; the importance of the rape motif, power relations between the races realised through power relations between bodies, black men overpowering white women's bodies in rape, white men overpowering black men's bodies in castration and lynching. The ideology of the very notion of race invites these narratives; race is an idea in the discourse of biology, a way of grouping people according to perceptions of bodily difference.

Yet the problem of the body seems to me to be rooted not in the biologism of race so much as in the justification of the capitalist system itself. The rhetoric of capitalism insists that it is capital that makes things happen; capital has the magic property of growing, stimulating. What this conceals is the fact that it is human labour and, in the last instance, the labour of the body, that makes things happen. The body is a 'problem' because to recognise it fully would be to recognise it as the foundation of economic life; how we use and organise the capacities of our bodies *is* how we produce and reproduce life itself. Much of the cultural history of the past few centuries has been concerned with finding ways of making sense of the body, while disguising the fact that its predominant use has been as the labour of the majority in the interests of the few. One way of doing this has been the idea of sexuality, an ever increasing focus on the genitals as a concentrate of physical needs and desire. Another has been the professionalism of medicine and the medicalisation of ever increasing aspects of bodily function, notably those connected with the reproduction of life. Yet another has been race, which at the level of representation means blacks, since whites are represented not for the most part *as* whites but as the human norm.

It is no accident that blacks should figure so crucially in this scheme of things. Through slavery and imperialism, black people have been the social group most clearly identified by and exploited for their bodily labour. Blacks thus became the most vivid reminders of the human body as labour in a society busily denying it. Representations of blacks then function as the site of *remembering and denying* the inescapability of the body in the economy. Hence, on the one hand, the black body as a reminder of what the body can do, its vitality, its strength, its sensuousness; and yet, simultaneously, the denial of all that bodily energy and delight as creative and productive, seen rather hysterically in images of bad (mixed) blood and rape or else as mere animal capacity incapable of producing civilisation.

Hence, finally, a figure like Robeson, whose body can be, in sport, in feats of strength, in sculpturable muscularity, in sheer presence, in a voice the correlative of manly power, but whose body finally does nothing, contained by frames, montage, narrative, direction, vocal restraint. He was a cross-over star because (and as long as) he so hugely em-body-ed, in-corpo-rated this historical functioning of black people in Western representation and economy.

3:

Judy Garland and Gay Men

He once told me about picking up a bloke who said you could always tell a 'queer's' place because they've all got LPs of Judy Garland.

Kenneth Williams, speaking of Joe Orton

The white kids had the counter-culture, rock stars and mysticism. The blacks had a slogan which said they were beautiful, and a party demanding power. Middle America had what it always had: Middle America. The hawks had Vietnam, and the doves the Peace Movement. The students had campus politics, and the New Left had Cuba and the Third World. And women had a voice. I had rejection from all of them. I also had Judy Garland.

Drag Queen in **As Time Goes By**

In the June 1973 issue of the Birmingham Gay Liberation Front newsletter, there was an article about Judy Garland. It was called 'Born in a Trunk' and was printed on pink paper. It made no reference to gayness whatsoever. The author did not feel that there was any need to explain why there should be a straightforward fan's account of Garland's life in the publication of a militant gay political group. Nor did the editorial collective, of which I was a member (though I neither wrote nor suggested the article). We had a policy of printing anything anybody in GLF submitted provided that it was not sexist, racist or fascist. The Garland piece was none of these, and we were as immersed in gay male culture as the author so that printing a piece on Garland seemed like the most natural thing in the world. When we distributed it however, a number of people, gay and straight, asked us what on earth the piece was doing in the newsletter. A few were objecting on the grounds that Garland belonged to the unliberated days of gay existence before GLF, but the rest were mystified. Why was Garland in a gay magazine? And when they knew the answer – because so many gay men (especially) are into her – the next question was a a bemused 'why?' This chapter is an attempt to answer that question.

There will not, of course, be one answer, but a variety of ways in which a star's image can be read if it is to attain star-sized currency and appeal. Hence Monroe can be read within different discourses on sexuality and similarly in those other than sexuality. Robeson too had to appeal across white and black audiences, and with far greater differentiation within them than I have described. Similarly, not only did Garland have the requisite massive appeal to non-gay audiences, but various aspects of her image spoke to different elements within male gay subcultures. In this chapter I want to explore how specific aspects of Garland's image could make a particular set of senses for gay men.

What I am describing is specific, both in terms of the subculture referred to and in terms of period. The subculture is particular first of all by being male. Historically lesbian and gay male subcultures have been linked, though often only tangentially; equally, lesbian subcultures have used Hollywood stars as important figures in their discourse (see Meyers, 1976; Sheldon, 1977; Whitaker, 1981; Gramann and Schlüpmann, 1981). However, lesbian and gay male cultures are not one, and I have no sense of Judy Garland being an especially significant figure in the lesbian subcultures. Greta Garbo and Marlene Dietrich, on the other hand, have been important for both subcultures, and it would be instructive to draw out the links between their images and the aristocratic dyke culture associated with Radclyffe Hall, Romaine Brooks, Lady Troubridge, Gertrude Stein and others (see Ruehl, 1982 and Dyer, 1983). Though this dyke style would be an important reference point for the gay male readings of Garbo and Dietrich, they would still need to be distinguished from lesbian readings.

The relevant male gay culture is further particularised by being urban (indeed usually metropolitan) and white. This does not mean that small-town, provincial and non-white gay men could not share it, but that it was produced in the developing urban gay male ghettoes (New York, London, San Francisco, Amsterdam, Sydney etc.) and fostered in forms (drag shows, bars) and publications largely controlled by whites. Urban white gay men set the pace for this culture, and in the period under consideration largely defined it as gay male culture itself (see Cohen and Dyer, 1980).

The period under consideration occurs after 1950. It was in that year that Judy Garland was sacked by MGM and tried (rather more desultorily than the press allowed) to commit suicide. This event, because it constituted for the public a sudden break with Garland's uncomplicated and ordinary MGM image, made possible a reading of Garland as having a special relationship to suffering, ordinariness, normality, and it is this relationship that structures much of the gay reading of Garland. In part this reading focuses on her subsequent career – the development of her concert appearances beginning at the London Palladium in 1951; her vehicle films, **A Star is Born** (1954) and **I Could Go On Singing** (1962), as well as starring in **A Child Is Waiting** (1962) and having a showy cameo in **Judgement at Nuremberg** (1961); a series of albums for Capitol records between 1955 and 1965, notably **Miss Show Business, Alone, The Letter, Judy in Love** and the double album **Judy at Carnegie Hall** (a recording of her 1961 concert that was the first double album to sell over a million copies); television shows, including two successful specials (**The Ford Star Jubilee** 1955, **The Judy Garland Show** 1963) and a less successful series for CBS in 1963; as well as interviews, radio shows, TV chat show appearances and much press coverage, chiefly of the vicissitudes of her life (suicide attempts, divorces, hospitalisations, and the like).

The post-1950 reading was also a reading of her career before 1950, a reading back into the earlier films, recordings and biography in the light of later years. This was facilitated by the growth of television and, in large cities, of repertory cinemas specialising in nostalgia revivals, both making Garland's films constantly available for reviewing. In addition, much of her post-1950 career deliberately evokes and reworks the early career. Both **A Star is Born** and **I Could Go On Singing** are clearly based on Garland's life story, and *Born in a Trunk* in the former is like a knowing précis of the image MGM had fostered (cf. Jennings, 1979). Equally her concerts were built around her film career, introducing only a limited amount of new material. Medleys of past hits were introduced with verses such as 'The story of my life is in my songs'. As Christopher Finch (1975, p.186) puts it, her concerts were

> *a novel kind of Broadway musical, the words and music by various writers and composers, the book by Judy Garland, with the formidable assistance of the entire Hollywood press corps. To the audience, the book was Judy Garland's life story.*

Equally important, during this period Garland spoke increasingly about her life before 1950 (notably in an article in **McCall's** in 1952), thus providing still more opportunity of reading those films and images through later knowledge and understandings.

Because of the availability of the earlier work and its

importance as a reference point in the later films, concerts, records and press coverage, the whole of Garland's career is relevant to this chapter, but it is read through the way she was taken up by the gay male subculture after 1950. The fact of the importance of Garland to this subculture was always widely noted. Most of the obituaries mention it, and it was particularly noticeable, for straight observers especially, at the concerts. Al Di Orio Jr (1975, pp.133–4) quotes from the **Los Angeles Citizen News** review of her 1961 Hollywood Bowl concert:

> *They were all there, the guys and dolls and the 'sixth man', sitting in the drizzle which continued throughout the concert . . . After 'Over the Rainbow' the standing, water-soaked audience applauded until Judy came back and sang three more songs. The guys and gals and 'sixth man' wanted more.*

'The sixth man' is a reference to Kinsey's findings on the incidence of homosexuality in the American male, a statistic familiar enough to provide the title for a sympathetic exposé of homosexuality by Jess Stearn in 1962. Non-gay observers are often more venomous than the rather neutral **Los Angeles Citizen News** reporter. William Goldman's (1969, pp.3–4) description of the last night of Garland's 1967 Palace season displays his obvious straightness – a beautiful gay man is a shock to him. He quotes without comment a straight man's reference to Auschwitz, as if the Nazi extermination of gays is of no account, and he includes the same man's gag, presumably for the reader to laugh at; all this laced with the usual vocabulary of homophobia, 'boy', 'flit', 'chatter', 'oooh', 'flutter', 'fags':

> *as the lobby filled up entirely, the audience itself began to become insistently noticeable. A stunning blonde walked by, in a lovely green jacket, sexy and confident, undulating with every step, and it comes as a genuine shock to realize the blonde is a boy. Two other boys flit by, chattering. First: 'I got her pink roses and white carnations; you think she'll like it?' Second (angry): 'Now why didn't I bring her flowers? Oooh, it's just too late for me now.' Another flutter of fags, half a dozen this time, and watching it all from a corner, two heterosexual married couples. 'These fags,' the first man says. 'It's like Auschwitz – some of them died along the way but a lot got here anyhow.' He turns to the other husband and shrugs. 'Tonight, no one goes to the bathroom'.*

I do not intend to go any further into these straight accounts of gay men and Judy Garland. I mention them here partly to indicate how widely observed the gay-Garland connection was and partly to register the extent to which this was felt as in some way offensive or threatening, an index of the degree to which gay men's use of Garland's image constituted a kind of going public or coming out before the emergence of gay liberationist politics (in which coming out was a key confrontationist tactic). This in fact is how going to a Garland concert in Nottingham in 1960 is remembered by one gay man:

> *I shall never forget walking into the Montfort Hall. Our seats were very near the front and we had to walk all the way down the centre gangway of a hall already crowded. I should think every queen in the east Midlands catchment area had made it . . . everyone had put on their Sunday best, had hair cuts and bought new ties. There was an exuberance, a liveliness, a community of feeling which was quite new to me and probably quite rare anyway then. It was as if the fact that we had gathered to see Garland gave us permission to be gay in public for once* (Letter to author).

I'm going to begin by looking at what gay men have written about Judy Garland and then move on to consider her image in relation to general aspects of the gay male subculture. There will be some overlap – the gay writings stress Garland's emotionality and its relation to the situation of gay men, and I'll relate this to aspects of her image and performance; equally when discussing those aspects of her image that suggest a connection with gay culture – ordinariness, androgyny, camp – I'll refer where possible to other gay writers' observations on these aspects. The difference in the two sections is of degree – the first is concerned with what is characteristically and explicitly referred to in gay writings, the second to what is more evident by putting her image side by side with characteristic features of gay culture in general.

By gay writings I'm referring equally to articles in gay publications and to responses I received to letters I placed in gay newspapers and magazines.* In terms of content there is no significant difference between these two kinds of writing. What does need to be pointed out is that all were written after Garland's death and the emergence of the modern gay movement. Many have memories of the earlier period, but are often articulated in terms that may well have been clearer later – for instance, I do not doubt the memory of the writer, quoted above, who speaks of Garland's Nottingham concert giving 'us permission to be gay in public for once', but equally to articulate things in terms of the

*Namely, **Gay News** and **Him** (Britain), **Body Politic** (Canada, **New York Native** and **The Advocate** (USA). These range from the broadly political/social papers like **Gay News** to more pornographic magazines like **Him.** The letter asked people to write to me about their memories of Judy Garland and why they liked her.

importance of being 'gay in public' is very gay liberationist in emphasis. The writings are all informed by an awareness of gay politics, though there is an interesting shift between some of the earlier writings and some of the later. This is signalled by a change in the pronoun used to refer to gay people, from 'they' to 'us'. This goes along with a changed emphasis from Garland representing gay men's neurosis and hysteria to her representing gay men's resilience in the face of oppression. Thus Roger Woodcock (1969, p.17), in the first issue of **Jeremy** (a very softly pornographic and, initially, an ostensibly bisexual publication, which appeared just before the development of the gay liberation movement in Britain), wrote:

> *Every time she sang, she poured out her troubles. Life had beaten her up and it showed. That is what attracted homosexuals to her. She created hysteria for them.*

Barry Conley (1972, p.11), writing in an early **Gay News** (which had been started under the impetus of the British gay liberation movement), puts a more positive emphasis ('fighting back') but still refers to 'they' and makes a comparison with straight people to the detriment of gay:

> *She began to gather a large following of homosexuals at her concerts, who were eager to applaud each and every thing she did . . . Perhaps the majority of those audiences saw in Judy a loser who was fighting back at life, and they could themselves draw a parallel to this . . . One should also remember that she still managed to retain all her 'straight' admirers over the years, though of course these people were less exhibitionistic in their reactions to her concerts.*

Nine years later, Dumont Howard (1981, p.95) in **Blueboy** (a gay male equivalent of both **Playboy** and **Cosmopolitan,** with that odd mixture of civil rights editorialising and consumer/hedonist copy, plus sexy pictures) took issue with the reading of gay men's enthusiasm for Garland as 'hysterical' and 'exhibitionistic', at the same time himself owning the positive feelings he was describing by the use of the pronoun 'we':

> *Garland is often painted as a pathetic figure and her fans – particularly her gay fans – as devotees of disaster . . . Now that we, as gay people, are learning more about ourselves as a group and a culture, we can begin to understand the true attraction of Judy Garland: it is her indomitable spirit, not her self-destructive tendencies, that appeals to gay audiences.*

Similarly, in the Gay Sweatshop production **As Time Goes By** (a play concerned with the recovery of the radical moments in gay history of the past century), in the final section (a series of interweaving monologues in a New York gay bar in June 1969),* the drag queen, hauntingly played in the original production by Drew Griffiths, says:

*The emergence of the contemporary gay movement is commonly dated from an incident at the Stonewall bar in New York in June 1969, also the month and year in which Judy Garland died.

> *They say we loved her because she mirrored the anguish and loneliness of our own lives. Crap. My parents were straight . . . They were the most anguished and lonely people I ever knew. No. We do not have a monopoly in the anguish and loneliness department. I loved her because no matter how they put her down, she survived. When they said she couldn't sing; when they said she was drunk; when they said she was drugged; when they said she couldn't keep a man . . . When they said she was fat; when they said she was thin; when they said she'd fallen flat on her face. People are falling on their faces every day. She got up* (Greig and Griffiths, 1981, p.62).

The interpretation in the early writings of what gays saw in Garland is similar to many of the unpleasantly homophobic observations of critics, particularly in the obituaries. One might say that Woodcock and Conley are victims of a kind of self-oppressive 'false consciousness', internalising straight interpretations of the gay response to Garland. Maybe, but it would be wrong to assume that only the more 'positive' readings accurately express the range of ways gay men might take the Garland image. Woodcock and Conley condense several levels of self-oppression which may in fact characterise one way of reading Garland. They are gay fans of Garland who distance themselves from any gay identification ('they', 'homosexuals') by putting down gay responses to Garland. At the same time, they can only read Garland in a gay way that is negative ('hysteria', 'exhibitionistic') – they are recognising a quality of emotional intensity that is in fact what the other gay writers also emphasise but they give it a denigratory label. Later readings want to disown all this, quite properly perhaps – but Garland could also be used in this subcultural discourse, more queer than gay, that spoke of the homosexual identity in self-oppressive modes characteristic of oppressed groups – distancing, denying, denigrating. Aspects of Garland's career and performance could be seen as pathetic and God-awful, and gay men could as it were misrecognise themselves in that and hasten, as Woodcock and Conley do, to disown it. I mention this here because they are the only writers to speak of Garland like this, but are also those writing from their most immediate memories of her. In sharing the predominantly positive understandings of what gay men saw in Garland of the other writings, I don't want to discount altogether this negative reading which the gay liberationist context of the later writing may have filtered out.

There is one source of references to Garland written before 1969, but it is difficult to handle. The British film monthly **Films and Filming,** which began in October 1954, quickly established itself as a closet gay magazine. It consistently published pin-ups of half-clothed male stars and starlets, advertised the kind of fashionable/sexy clothes shops patronised by gay men, and ran a lively small ads column that by the sixties had become quite explicit in its correspondence section ('Gay young man wishes to meet another, varied interests including physiques', November, 1964, p.31). In **Films and Filming** the gay-Garland connection is clear, if never explicitly referred to. Ads for the Garland fan club appeared every month (and this was not so for any other star) amid ads for physique films and 'discreet bachelor apartments'. One small ad made the connection as directly as was possible:

> *Young man leaving London for Durham in early October seeks similar (age 21–23) living in that area – view to writing initially, meeting soon. Interests, films, music (serious), Judy Garland, photography, driving (own car). Photograph welcomed. Box 856F (***Films and Filming,** July, 1964).

There was a picture of Garland (on the set of **A Star is Born**) in the first issue and she was on the cover of the third (December 1954), a still of *The Man that Got Away* from **A Star is Born.** Inside there is a long celebratory article by the magazine's editor, Peter Brinson, which, while it does not make any reference to her gay following (which can anyway only have been embryonic then), does write of her positive qualities. I'll refer later to this article; I've brought it in here because it is the one piece of rather fragile evidence from the pre-gay movement period about how gays made sense of Judy Garland, and it is a positive one. Just as I don't want the post-gay liberation readings to drown out the negative, self-oppressive readings, so equally I don't want to give the impression that the more positive readings were only possible after the gay movement had started. On the contrary, the only evidence we have suggests the positive reading was always predominant.

The common emphasis in all gay writings on Garland is on her emotional quality. As Dumont Howard (1981, p.95), puts it:

> *The essence of Judy Garland's art is emotion. She burns right through lyrics, delivering, instead, their pure emotional substance.*

Many of the accounts do not refer to the gayness of the emotionality, but rather suggest the immediate, vivid, intense experience of it. This is worth stressing because however much one can see that Garland is appreciated in ways specifically relating to gay culture, she is not necessarily experienced like that. As with any star, the fan's enthusiasm is based on feeling that the star just is wonderful. Thus some of the people who replied to my letter in the gay press said that as gay men they liked Garland because she had, for instance, 'star quality' or 'great talent and warmth', terms that might be used by an enthusiast for any star. The intensity and excitement of our experience of them outstrips our consciousness of what they stand for.

The kind of emotion Garland expressed is somewhat differently described in the gay writings, but on two points all agree – that it is always strong emotion, and that it is really felt by the star herself and shared with the audience.

> *In every song I've heard it gives you something of her as a person, all the tragedy and happiness of her life is echoed in every word she sings* (Letter to author).

> *All the emotions felt inside worn on the outside. The voice a clarion call of power, joy, love, sharing it all with us* (Letter to author).

Although these are qualities that might be attributed to many stars, it is the particular register of intense, authentic feeling that is important here, a combination of strength and suffering, and precisely the one in the face of the other. Different writers put different stresses on just how the two elements are combined. Some see the strength all but denying the imputed suffering:

> *Judy . . . the power, the strength, the defiance of all the washed-up-has-been talk* (Letter to author).

Peter Brinson's (1954, p.4) article in the third issue of **Films and Filming** details MGM's exploitation of her and other personal setbacks, but the keynote is the celebration of her strength:

> *One kind of courage I admire is that which keeps going, come what may. Judy Garland is an example.*

Others emphasise her openness to suffering, her ability to convey the experience of it:

> *gay people could relate to her in the problems she had on and off stage* (Letter to author).

> it was precisely the quality that was the cause of all the pain that was also appealing to her audience. When she sang she was vulnerable. There was a hurt in her voice that most other singers don't have (Bronski, 1978, p.202).

The least attractive expression of this emphasis, with ugly photos as accompaniment, is Kenneth Anger's (1981, pp.413–6) reference to Garland in **Hollywood Babylon,** written after her death and even here recognising the strength while dwelling on the point at which it gave out:

> *MGM's Amphetamine Annie really made it at last after so many attempts – pills, wrist slashings years before in her Hollywood bathroom, hack hack with broken glass . . . She was* **hundreds** *of years old, if you count emotional years, the toll they take, dramas galore for a dozen lifetimes. She was 'She' who had stepped into the Flame once too often.*

Wherever the emphasis comes it is always the one in relation to the other, the strength inspirational because of the pressure of suffering behind it, the suffering keen because it has been stood up to so bravely.

Films and Filming's article is headed 'The Great Come-Back' and the come-back was the defining motif of the register of feeling I'm trying to characterise, for it is always having come back from something (sufferings and tribulations) and always keeping on coming, no matter what. Repeatedly in the news from 1950 onwards for this or that reason (suicide attemps, failed marriages, drunk and disorderly charges, and so on), Garland repeatedly came back (Oscar nomination for **A Star is Born,** most successful ever double album **Judy at Carnegie Hall,** sell-out at all concerts and cabaret performances). The very act of coming back set off the feeling and it was reprised in countless details.

Both **A Star is Born** and **I Could Go On Singing** end with the Garland character coming back from personal despair (widowhood and rejection by her child, respectively) to a public performance. The latter has her sing at this come-back performance the come-back song that is the title of the film. The opening number of the 1961 concert (and record) is *When You're Smiling,* itself a 'keep on keeping on' song, with a bridge passage tailor-made for her image. It begins with rueful references to the kinds of problems one should smile in the face of, problems very close to those of Garland herself (marriage, weight, drugs):

If you suddenly find out you've been deceived
Don't get peeved
If your husband bluntly tells you you're too stout
Don't you pout
And for heaven's sakes retain a calm demeanour
When a cop walks up and hands you a subpoena
If the groom should take a powder while you're marching down the aisle
Don't weep and moan
Because he's flown
Just face the world and smile.

Then as she leads back into the melody, she could be making two references quite specific to herself:

'Cos when you're crying don't you know that your make-up starts to run
And your eyes get red and scrappy?
Forget your troubles, have yourself a little fun,
Have a ball
Forget 'em all
Forget your troubles, c'mon, get happy
Keep on smiling
'Cos when you're smiling
The whole world smiles with you.

First, she sings about make-up. In part this is singing about her immediate situation, made up for a performance that was also a come-back. The song itself is a vaudeville standard, associated with Al Jolson, the epitome of the showbiz ethos, with whom Garland, dubbed in the fifties 'Miss Show Business', was often compared. It evokes a whole showbiz litany of tears-beneath-the-greasepaint, the show-must-go-on, that gives a resonance of tradition to Garland's coming back. But in addition one of the most frequently repeated stories about Garland was how she used the act of preparing herself for public appearances as an answer to problems. For example, Roger Woodcock (1969, p.16) in **Jeremy** writes:

She knew they said she drank too much and took too many pills and it upset her. 'What do you do when people talk about you like that?' she asks, 'commit suicide? No, that's messy. Get drunk? No, that's no solution.' What Judy did in fact was to put on her lipstick, make sure her stockings were straight, then she marched onto a stage somewhere and sang her heart out.

The make-up reference in *When You're Smiling* suggests the moment of pulling herself together by putting on her make-up.

Secondly, she interpolates a phrase from another smile-through-your-tears song, *Get Happy* – 'Forget your troubles, c'mon, get happy'. This was the final number in **Summer Stock** (1950), her last MGM picture. The reference not only reminds the audience of one of the cult numbers from her films, but more generally of her MGM image. It was also widely known, by 1961, that the making of **Summer Stock** had been fraught with difficulties of all kinds, including her perennial weight problems, and that *Get Happy* was in fact shot several weeks after the completion of the rest of the film, as an afterthought. Garland came back for it, much relaxed and rested, noticeably slimmer and rehearsing and shooting in one day a number that is, as Jane Feuer (1982, p.20) puts it, 'the ultimate in professional entertainment'. The facts about coming back to film *Get Happy,* the image of a confident, slimmer Garland on screen, as well as its lyrics, all encapsulate the come-back motif, and are condensed in the snatch of it included in the come-back song *When You're Smiling* that starts her come-back concert at Carnegie Hall.

This come-back, going on going on, suffering and strength quality could even be read in the performance of the songs, especially towards the end of her career. In the later concerts, the sense of the trials of her life was no longer offstage, in the publicity read beforehand, in what the songs and gags referred to, but could be seen in the frailty of her figure, heard in her shortness of breath and shaky high notes, noted in her lateness or stumbling walk. Yet she was still carrying on with the show. However demanding the melody now seemed for her, she did get to the end of the song and this became a mini enactment of the come-back motif. Al Di Orio (1975, p.201) quotes from the **Camden Courier Post** of 21 June 1968 on her delivery of *Over the Rainbow* at the Kennedy Stadium concert in Philadelphia two nights earlier. She speaks/sings:

If happy little bluebirds fly, beyond the rainbow, why – I made it, I made it – why, oh why – thank you, darlings, I made it all the way through, I didn't think I would – oh why, can't I?

Di Orio argues that this is a misquotation:

> *What she said was, 'I finally made it over the rainbow thanks to you all'. Then she continued with the song, then she yelled, 'We all do it, you know'. Then another phrase of the song, and finally, 'Thank you. God bless you'.* (ibid.).

I do not know whose memory of her exact words is correct, but may it not be that in the concerts getting to the end of the song enacted getting to the end of the rainbow? In her last recording, of a performance at the Talk of the Town in London, her apparent difficulty in getting through *Over the Rainbow* nonetheless ends with her coming through loud and true on the final 'I', producing a triumphant last note in the teeth of the ravaged voice that precedes it.

Gay writing returns repeatedly to this emotional quality as in some way representing the situation and experience of being gay in a homophobic society:

> *[They] saw in Judy a loser who was fighting back at life, and they could themselves draw a parallel to this* (Conley, 1972, p.11).

> *To the gay male, in those days, there was a terrific bond between Miss Garland and her audience, we, the gay people could identify with her . . . could relate to her in the problems she had on and off stage* (Letter to author).

> *She appeals to me as a gay person . . . because she tended to sing songs which seem to echo all the doubts and trials of a gay man within an unaccepting social order. 'The Man That Got Away' could almost become the national anthem of gay men . . . Others too express 'our' desires as a minority. The great 'Over the Rainbow' suggests a perfect world in which even we could live without restricting our life-styles and songs like 'Get Happy' tend to express our ability to cope and get along with 'our lot', whatever happens* (Letter to author).

This gives us then the feel of the gay sense of Judy Garland. Other stars can suggest this quality (Billie Holiday, Edith Piaf, Shirley Bassey) though for none has the come-back been so decisive a motif. Equally, other groups in society carry on in the face of social stigmatisation and Garland did appeal in these terms to other people. Why the special felt affinity between *this* 'emotional' star and *this* oppressed group?

I want to discuss shortly three aspects of gay culture which are consonant with aspects of Garland's image, but before doing that I want to look at gay writing which links Garland's emotional quality to a general emotional quality of gay life, the idea of a 'gay sensibility'. The key to this lies in the particularity of gay people's situation, namely that we can 'pass for straight':

> *The experience of passing is often productive of a gay sensibility. It can, and often does, lead to a heightened awareness and appreciation for disguise, impersonation, the projection of personality, and the distinctions to be made between instinctive and theatrical behaviour* (Babuscio, 1977, p.45).

This awareness informs responses to Garland in different ways. Jack Babuscio argues that the sense of Garland performing herself, enacting her life on screen or stage, is a recognition of the theatricality of experience that the gay sensibility is attuned to. Vito Russo (1980–1, p.15), on the other hand, stresses the nerve and risk involved in living 'on the edge' between a stigmatised gay identity and a fragile straight front, evoked in the very act of Garland going on stage in the teeth of disaster:

> *I'm not sure that people know what it means any longer to watch a performer walk onto a stage stone cold and suddenly be absolutely brilliant. When Garland sang 'If Love Were All' or 'By Myself', her whole life was on that stage, and believe me, that's not nothing . . . That's why I'm so attracted to Garland. She had the guts to take the chance of dropping dead in front of ten thousand people. And won.*
> *Is that a particularly gay response to Garland? Perhaps. Gays take chances all the time in ways straights never do. We have traditionally been forced to put on one face for the world and another in private.*

What both Jack Babuscio and Vito Russo bring out is the way that the gay sensibility holds together qualities that are elsewhere felt as antithetical: theatricality and authenticity. Equally I'd want to suggest that the sensibility holds together intensity and irony, a fierce assertion of extreme feeling with a deprecating sense of its absurdity. This is a quality I find in a number of popular songs by gay writers – Cole Porter's *Just One of Those Things,* Noel Coward's *If Love Were All,* Lorenz Hart's *My Romance,* for instance. Garland seemed to have a particular affinity for songs like this. Her version of the Haymes and Brandt song *That's All* (on **Just for Openers,** a collection of songs from her TV series) is a good example. The song uses the extravagant rhetoric of forever-love and is ironically off hand – 'I can only give you love that lasts forever, that's all'. In the final verse Garland gives a characteristic, all-out, slurred, Jolsoney delivery of the lyrics' demand for absolute love in return, then gives a tiny laugh before the final 'that's all':

If you're asking in return what I would want dear
You'll be glad to know that my requests are small
Say it's me that you'll adore
For now and evermore
(hm)
That's all
That's all.

I've spoken personally here, so I'd better make it clear that I'm not claiming that mine is the definitive gay response to Garland – I'm simply using myself in evidence alongside the writings used throughout the chapter. This passion-with-irony is another inflection of the gay sensibility, a doubleness which informs equally Russo's living-on-the-edge, Babuscio's theatricalisation-of-experience and indeed the whole suffering-and-strength motif.

Notions of a sensibility are elusive, though worth persevering with because they are attempts to get to grips with particular ways of feeling, something that semiotic criticism has shown little enthusiasm for. I want to turn now, however, to somewhat more concrete aspects of male gay culture, putting less stress on gay writing about Garland and more on bringing aspects of the culture in general together with aspects of Garland's image.

Garland works in an emotional register of great intensity which seems to bespeak equally suffering and survival, vulnerability and strength, theatricality and authenticity, passion and irony. In this she belongs to a tradition of women vocalists that includes Holiday, Piaf, Bassey, Barbra Streisand, Diana Ross (but not, say, Ella Fitzgerald or Peggy Lee), who have all been to varying degrees important in gay male culture. Garland's image may be identified the most with the gay male audience – Holiday was too early for any clear subcultural identification, Piaf was not a Hollywood star, and the others have not had to orchestrate the come-back as a symbol of survival as Garland did. But in terms of emotional register it is a question of circumstances and degree. Like all these women, she sings of desire for men and of relationships with men going wrong. Male singers could not (still largely do not) sing of these things. Gay men gravitated to women singers because they did sing of them, and perhaps also, for reasons discussed below, because gay men often thought of themselves as occupying a 'feminine position' by virtue of desiring other men. But Garland's very special place in this line of singers has to do with several other qualities of her image that were homologous with aspects of male gay culture. All the other singers mentioned have some of these qualities too but none has all of them, and none is thus so overdetermined as a potential gay men's star. The qualities I am referring to are ordinariness, androgyny and camp.

Ordinariness

It may seem surprising to suggest that ordinariness is part of the male gay reading of Garland. However banal gayness may in reality be, it is not usually thought of as being ordinary, and the urban gay male subculture (unlike the gay civil rights movement by and large) has never laid any emphasis on gays' ordinariness. It is not, in any case, Garland's direct embodiment of ordinariness that is important, the all-American, girl-next-door image that MGM promoted and that presumably accounted for a large part of her non-gay and pre-1950 appeal. What is important is rather a special relationship with ordinariness, particularly in the disparity between the image and the imputed real person, but also in the way she is set up in relation to movie going.

The insistent ordinariness of her MGM image is a prerequisite for the gay male reading. It cannot be overstressed just how dominant the image was, and we should also remember the degree to which ordinariness was offered as the ultimate moral attribute of the American way of life, rather than the ideals of piety, charity, heroism and so on, whose very idealness makes them not ordinary (cf. Marcuse, 1964). In a culture in which the images of the small town and next door are the touchstones of normal life, stories about girls who live in small towns and fall in love with boys next door become the epitome of ordinary life. This is the story of the majority of Garland's MGM films. The Andy Hardy films (**Love Finds Andy Hardy** 1938, **Andy Hardy Meets Debutante** 1940, **Life Begins for Andy Hardy** 1941) as well as those other films with Mickey Rooney that were all but Andy Hardy films (**Babes in Arms** 1939, **Strike Up the Band** 1940, **Babes on Broadway** 1941) were clearly packaged and understood as

At the soda fountain in **Babes in Arms**

Portrait of Judy Garland as tennis player: 'the small-town game'

Judy Garland in gingham in **Strike Up the Band**

Judy Garland in gingham in **Till the Clouds Roll By**

hymns to ordinariness, and promotion material on Garland and Rooney amplified this, showing them for instance at soda fountains (the innocent meeting place of small-town kids), playing tennis (the small town game) and singing at a simple upright piano with the stars and stripes on it. This small-town Americana iconography is repeated in their films together, most notably in the Andy Hardy series, themselves hymns to small-town America. The quintessential ordinariness of Dorothy in **The Wizard of Oz** (1939) (though an orphan, she lives on a small farm, another emblem of US ordinariness) is registered by her blue gingham frock, her costume equally in **Pigskin Parade** (1936), **Everybody Sing** (1938) **, Strike up the Band** (1940), **Little Nelly Kelly** (1940), **Till the Clouds Roll By** (1946). Some of the later films play on this. A chance encounter with a young man in New York, in **The Clock** (1945), leads to a series of incidents and vignettes culled straight from the imagery of the small-town films, rendering the anomic city into something more comfortable. **In the Good Old Summertime** (1949), a woman in love with a man she is writing to discovers that she works in the same shop as him – the discovery that your heart's desire is right where you are being the same structure as that of **The Wizard of Oz. Meet Me in St Louis** gives her the role and song (*The Boy Next Door*) that most warmly celebrates her ordinary image, although some recent critics have suggested that in all sorts of ways the film also undermines the very image of small-town family life it so apparently promotes (see Wood, 1976 and Britton, 1977/78; for a sophisticated analysis that remains close to the 'obvious' feeling of the film, see Bathrick, 1976).

In the most obvious way, Garland was the image of heterosexual family normality. How, as one of my students put it to me, was a group excluded from and oppressed by this normality able to turn Garland into such an identification figure? The ordinariness is a starting point because, like Judy Garland, gay men are brought up to be ordinary. One is not brought up gay; on the contrary, everything in the culture seems to work against it. Had Garland remained an image of ordinary normality, like June Allyson or Deanna Durbin (who proved her normality by leaving Hollywood and settling down, happily married), she would not have been so available as a gay icon. It was the fact, as became clear after 1950, that she was not after all the ordinary girl she appeared to be that suggested a relationship to ordinariness homologous with that of gay identity. To turn out not-ordinary after being saturated with the values of ordinariness structures Garland's career and the standard gay biography alike.

A sense of this structuring could only fall into place after 1950. The press coverage of the suicide attempt, the **McCall's** story of her tribulations at home and (which was nearly the same thing) at MGM suggested that beneath the happy gloss of normality in the MGM films (and the radio appearances, the pin-ups and records) there was a story of difference. In a moment of what would conventionally be considered bad style, the writer of the article on Garland in the Birmingham GLF newsletter remarks, 'On the *surface,* Judy's life was happy, but problems were *surfacing'* (my italics), exactly expressing the way in which a 'below-surface' became part of the surface of the image itself. Once this 'below-surface' was available, it became possible to look back at the films and discern, or think you could discern, not straightforward ordinariness but a special relationship to ordinariness.

In part this meant picking up on the camp elements in her image, which I discuss below, but equally and contradictorily as important are the 'authentic', 'natural' qualities always attributed to Garland (but not to other stars in the gay pantheon, Garbo, Hepburn, Davis, Bassey, Streisand, Ross). For the idea of a sense of difference below or within ordinariness resembles essentialist conceptions of homosexuality as a trait inborn or inbred (but in any event so early established as to be at least second nature), a trait that may be repressed but is always there. This notion of the natural and given quality of homosexuality has been equally important to the arguments of early gay rights reformers such as Edward Carpenter and Magnus Hirschfeld, to the Wolfenden committee and other post-war liberal discourses (including the Mattachine Society) and to the Gay Liberation movement (see Weeks, 1977 and Watney, 1980); and its affinity to the repression hypothesis favoured by **Playboy** and deconstructed by Michel Foucault (see the chapter on Monroe in this book) is evident. Such a view of an essential difference within the framework of being brought up ordinary requires a star whose image insists on her or his authenticity, since the 'difference' must be embodied as true and natural. (See Dyer, 1982 for further discussion of the notion of authenticity and Garland's relation to it.)

How might this essential difference be registered or discerned in Garland's image in her pre-1950 work? After that her difficulties (albeit overexaggerated or overpublicised) provide the keynote for her relationship to her early image. Her small part in **Judgement at Nuremberg** (1961) is about an ordinary hausfrau who reveals her extraordinary (for its pre-war time) friendship with a Jewish man; **I Could Go On Singing** (1962) has Garland (as rather obviously herself) trying to reinsert herself into family-and-marriage with her ex-lover and son, but the attempt clearly being shown as impossible. After 1950, the difficulty of the relationship to ordinariness could be expressed, but before?

In fact, the MGM films were never so uncomplicatedly normal. We caricature them when we see them thus, and take them at the level of both their own promotion and the condescension of contemporary reviewers. Although we don't know what people got from them, and although they do always end up on a note of affirmation of ordinariness, along the way they have dusted up a number of problems, contradictions, misfits, as any entertainment film must do if it is to connect with the lives of its viewers. What the gay reading-back into the film does is seize on those elements, become more conscious of them.

Two elements in particular of Garland's image in the MGM period are relevant here, emotional intensity and lack of glamour. I'll come back later to questions of the androgyny of her image and her camp humour, the former at odds with the sex role norms of the films, the latter tending to denaturalise their normality.

In the MGM films Garland plays a demure girl next door; even if she lives in New York, it is soon established that she comes from a small town and has small-town values (e.g. **The Clock** and **Easter Parade);** her hair is allowed to fall in a simple perm and she wears plain frocks or the kind of party clothes that nice young girls were supposed to. Yet when she sings, it is either in a loud, belting, peppy style or in a torch style. Her abilities as a belter were often stressed by comparing her with the more refined, parlour-singing style of, say, Deanna Durbin (in their MGM film – Garland's first, Durbin's only – **Every Sunday** 1936), or having her teach rising opera star José Iturbi how to swing it in **Thousands Cheer** (1943), or simply having her sing *Swing Mr Mendelsohn* in **Everybody Sing** (1938) (cf. 'The Judy Garland Opera vs. Swing Number' in Feuer, 1982, pp.57–9). Her singing of sad numbers in films of an otherwise happy-go-lucky tone – *I Cried for You* in **Babes in Arms** (1939), *But Not For Me* in **Girl Crazy** (1943), *Better Luck Next Time* in **Easter Parade** (1948) – introduces a note of pathos closer in intensity and emotional register to either blues or opera than any of the other material in these films. The roles as written, sold, dressed and directed are about containing the peppy emotions of the belting insubordination to Rooney or whoever else was the male lead, while the torchy numbers seem like sudden switches of register not picked up elsewhere in the film. The belting and the torchiness, both very much within registers of authenticity, are in excess of the safe, contained, small-town norms of the character, an essence of emotional difference akin to the idea of gayness as emotional difference born within normality.

The clearest elaboration on this is in **Meet Me in St Louis,** whose warm decor, harmonious scoring, comfortable cameo performances and glow of nostalgia beguile one into thinking of it as a straightforward evocation of the safety of small-town family life. The 'darker' elements do not come most strongly from Esther/Garland but from Tootie (Margaret O'Brien) and the Hallowe'en sequence; but the sense of the family containing Esther/Garland's pep and intensity is nonetheless present. As Andrew Britton has argued (Britton, 1977), the film both represents forcibly the strength and vitality of its women characters and yet has them put all that energy into securing their own subordination to a man, either by maintaining (Mrs Smith) or getting (the Smith sisters) a husband. In terms of narrative Esther/Garland is the Miss Fix-It of the household, whether mediating between the family on how sweet or sour the ketchup should be, or trying to arrange for Rose to have her telephone call with Warren in private, or standing up for Tootie when she thinks the latter has been beaten up by the boy next door, John (Tom Drake). In terms of numbers, she not only has peppy songs to sing like *Skip to My Lou* and *The Trolley Song,* but is shot and directed in such a way that the energy and flow of the song seem to come from her, either because she initiates the square dance patterns of the former or from the way the chorus groups around her and responds to her gestures in the latter. In these ways **Meet Me in St Louis** is a celebration of Garland's peppiness; yet the film also shows the awkwardness of this pep, the need to contain it. This is done humorously, but nonetheless consistently. Her peppiness makes her forward or leads her into embarrassing situations – it is she who sings the man's part to John Truitt in *Over the Bannister Leaning,* having carefully stage managed a romantic situation (turning low the lights); so fierce is she in defence of Tootie (mistakenly as it turns out) that John backs away from her when she returns to apologise, her actions having confirmed his inept observation at the end of the party that she has 'a mighty strong grip for a girl'. Many of the examples here I'm taking from Andrew Britton's analysis, and he notes the way that at the end of *The Trolley Song,* on the words 'with his hand holding mine', Esther/Garland takes hold of her own hand, a gesture that suggests how much the erotic charge of the story of the song emanates from her alone. This pep must be confined if she is to get the boy next door and end up like her mother – by pinching the 'bloom' out of her cheeks, by squeezing herself into the tightest corset, by accepting the humiliation of going with Grandad to the Christmas ball and dancing with the least attractive men. It is only after all this that John proposes to her.

Following the proposal, Esther/Garland goes home and finds Tootie crying at the thought of leaving St Louis for New York; she sings to her the one torchy number in the film (though the title of the song would not alert one to the intensity of the way it is sung and used in the film), *Have Yourself a Merry Little Christmas*. So affecting is the song that, far from being comforted, Tootie rushes out into the garden and hacks down the snow people she and her sisters and brother had made. Whether you take these snow people to represent parents or, as Tootie says, her elder siblings and their boy or girl friends, it is a violently emotional moment in which Tootie seeks to destroy the representatives of her social world. Yet the intensity that has got to Tootie may also be understood to spring from Esther/Garland's sense of defeat, her recognition that getting what she wants (John as husband) marks the end of pep and vitality. Certainly the placing of this sadly ironic song and the singing of it with such yearning intensity means that it expresses more than uncertainty about where they will all be next Christmas (which is all the lyrics would suggest).

In **Meet Me in St Louis** then Garland is the nice ordinary girl with extraordinary reserves of peppy energy and torchy emotion. In terms of the social values of its day, there is no reason why the gradual containment of this energy and emotion should be regretted – growing up was about containing feeling; getting a husband was what adolescent female energies were for. But given the vagueness and almost comic straightness of John (that first shot of him standing with his pipe in the next door garden) and given the pallid (blue and cream after all those reds, browns and yellows) and unspectacular quality of the final scene (which one would expect to be a splurge of spectacle in an MGM musical) one could certainly come away remembering best the pep and torchiness, the emotional difference contained within the normal confines of the Smith family.

Esther is clearly meant to be a pretty girl and Garland is shot in the film in a conventionally glamorous way – in **Meet Me in St Louis** we are supposed to think of Judy Garland as physically attractive. But in many ways this is unusual; more often there is a clear indication that we are not to think of her on a par with other

Judy Garland singing to a picture of Mickey Rooney in **Babes in Arms**

Dear Mr Gable (You Made Me Love You) from **Broadway Melody of 1938**

Hollywood female glamour stars. Her sense of inadequacy and inferiority as compared to these stars of her own age and period was recorded in the 1952 **McCall's** article and elsewhere. In the light of this one can get a strong sense of it being structured into her image and films. In **Everybody Sing** Billie Burke, playing Garland's mother, refers to her as her 'ugly duckling' and this was to have been the title of the film. Later Burke says to her and her (film) sister, 'You're looking very pretty this morning, both of you, even Judy.' When Allan Jones introduces her first appearance at the night club he tells the audience they are about to see a 'real prima donna' and makes the well-known gesture indicating a curvaceous female anatomy – it is supposed then to be cute and funny when dumpy little Judy comes on to sing. During *I Cried For You* in **Babes in Arms,** she (as Patsy Burton) soliloquises about herself:

> *I know I'm no glamour girl . . . But maybe someday you'll realize that glamour isn't the only thing in this world . . . Anyway, I might be pretty good-looking myself when I grow out of this ugly duckling stage. And you're no Clark Gable yourself.*

She is sitting looking at a picture of Mickey Moran/Mickey Rooney in a pose reminiscent of that for *Dear Mr Gable (You Made Me Love You)* in **Broadway Melody of 1938,** the embodiment of the girl sick with love for a man she cannot possibly attain. Rooney never has a number or speech similar to this in any film, and both films and promotion suggested he was constantly pursued by girls. Garland is not even as attractive as her leading man.

She is often compared to other glamour girls in the films – with Lana Turner in **Love Finds Andy Hardy** (and the threat of losing Andy Hardy to a glamour girl seems to be a recurrent narrative element in the series), to June Preisser in **Babes in Arms,** to Ann Miller in **Easter Parade,** to Gloria de Haven in **Summer Stock.** (In **Easter Parade** Miller/Nadine can make men's heads turn simply by walking down the street, whereas Garland/Hannah has to pull a grotesque funny face to achieve the same effect.) **Ziegfeld Girl** (1941), a film very consciously using the images of its three female stars, compares her to both Lana Turner/Sheila and Hedy Lamarr/Sandra as prospective Ziegfeld girls. Lamarr/Sandra most approximates to the Ziegfeld model of female pulchritude, but her very classiness makes her reject the Follies by the end of the film. Turner/Sheila is closer in spirit to the burlesque sexual display of the Ziegfeld enterprise, but her very vulgarity makes her unstable, unable to be truly professional, and

Ziegfeld Girl (1941): (l. r.) Hedy Lamarr, Judy Garland, Lana Turner

her career ends in tragedy (see Dyer, 1977–8). Garland/Susan is the only one to survive, but she does so because of her professional, born-in-a-trunk skills and talent, not because of her Ziegfeldesque looks. In the big production numbers, it seems that the film does not quite know what to do with her. In a film of admittedly grotesque costuming, there is still a fit between what Lamarr/Sandra and Turner/Sheila wear and their particular glamour image. In the production still showing the costumes for *Minnie from Trinidad,* Lamarr/Sandra wears a costume that covers her body and emphasises her height (the 'statuesque' quality supposedly characteristic of the Ziegfeld Girl) – her breasts and vulva are suggested by the placing of the outsize flowers, but the fact that these are orchids and other exotic-looking flowers draws from an iconography of class and wealth, the world of haute couture that Ziegfeld shows often called upon. She stands straight with an air of indifference on her face. Turner/Sheila, on the other hand, stands with one of her legs presented to the viewer, with bare arms and a more sultry expression on her face, as befits her more directly sexual image (as both character and star). The costume emphasises her body, the flowers at her bosom not, as with Lamarr/Sandar, symbolising the 'beauty' of her breasts but drawing attention to their actual shape beneath the fabric. Garland/Susan, on the other hand, wears a skirt that is open to above the knees and her shoulders are bare, but the costume neither shows off her body nor symbolises it – on the contrary, the conflicting lines of the fabric itself, the way it hangs and the accessories of loose belt and bow all cut across and break up her body shape. Though all the costumes are from a certain perspective ridiculous, Garland/Susan's hat here is surely ridiculous even in the costuming's own terms. She stands simply and smiles frankly, neither haughty nor sexy. The film does not in dialogue refer to Garland/Susan's looks, but both narrative (she gets to the top on talent) and costuming suggest that she does not have and cannot carry off glamour.

Not being glamorous is to fail at femininity, to fail at one's sex role. She might be valued for her peppy singing, but pretty much as one of the boys. Lack of glamour – and the painful sense of this registered in the torchy numbers which are often occasioned by a sense of inferiority – might correspond with several different ways in which gay men think about themselves in relation to their sexuality and sexual attractiveness, as gender misfits (see below), as physically deformed (if not bodily, at least biologically), and so on. But of course Judy does get Mickey, does get to sing for Clark Gable at his birthday party, does get to be the star of a Ziegfeld show. The pleasure of identification with this misfit could also be that she does get her heart's desire, as in wish fulfilment alongside her we may too.

There is one further important aspect of the way that Garland is situated in relation to ordinariness; as a movie fan. The moment that established her as a star was the image of her singing to a photograph of Clark Gable; early promotion photos showed her as an avid moviegoer. She is not so much a movie star herself as a stand-in for us the audience. In all her MGM films this is the point of departure, and also the point of entry for us into the magic world of films, for the Garland character is with us outside the magic movie world and yet has the ability to enter it. The films with Rooney play on this in relation to fan worship. He is always the star of the small town, the boy all the girls adore, the leader of the band; Garland is the plain girl who longs for him and gets him in the end. Other films place her outside the surrogate world of movie magic – Oz, glamour in **Ziegfeld Girl,** show business in **Presenting Lily Mars, Easter Parade, The Pirate** and **Summer Stock.** We go to see the films, for the spectacle of Oz, of big production numbers, of romantic moments. Garland is the person initially outside all this and as taken with it as we are; she enters the spectacle and becomes part of it, as we in our absorption in the dark may.

MGM films both celebrate ordinariness and offer the most extravagant spectacle and roster of stars of all the Hollywood studios; Garland makes the link between these two aspects of the studio's product, makes the spectacle accessible to ordinariness. She is able to articulate directly the desire to escape into the world of the movies. The use of movies as escape is a defining feature of the construction of entertainment, which has had a particular importance in gay men's lives. The isolation of the movie house from other forms of social interaction (other, of course, than heterosexual petting) and the felt isolation of gays from their heterosexual peers are of a piece, giving the movies a very particular place in the experience of many gay men. Garland can be the representation of the desire to live in the movies, because she is herself an ordinary person who escapes into the magic of films.

The concerts amplify this by their constant reference back to her old films; the more she could be treated as 'one of us', the more alive could be the sense of her relationship to the movies, one of us who entered the magic world. Only now it is nostalgia for having entered that world. Part of the intensity of *Over the Rainbow* at the concerts is a meeting of the moment of yearning to enter the magic world and the moment of remembering entering it. It is yearning for the moment of yearning, perhaps a desire to recover the innocence of believing you could enter the magic world. Just as Garland lost her innocence in her subsequent life (as it was publically known), so gay men could construct their own biographies as a loss of innocence as dramatic as hers. We do well to remember, of course, that the notion of the innocence of young people's relation to fiction is itself a culturally and historically specific understanding, and one that probably structures most nineteenth and twentieth-century biography. Gay men are not alone in feeling they have lost their innocence – Garland doing *Over the Rainbow* only takes on a particular resonance for the gay reading because it is of a piece with all the other aspects of her image that can be read from a gay perspective.

Androgyny

If Garland could be made to articulate a relationship to ordinariness, that was also because she could be held to be 'different'. In particular, she could be seen as in some sense androgynous, as a gender in-between. It is common in nineteenth and twentieth-century thought to conflate sexuality and gender. Biological sex difference is assumed to give rise to different sexualities; biological sex gives not only the social sex roles of masculinity and femininity, but also 'appropriate' roles for men and women within heterosexuality. In this context, homosexuality is viewed as 'in-between', an absence of heterosexuality which must go along with an absence of true or full masculinity or femininity. Hence gay men and lesbians are in-betweens, androgynous by gender because not fitting the masculinity and femininity bestowed by heterosexual sexuality. It is important to stress that this concept of a biological basis to homosexuality has not only been used against gay people (gays seen as deformities of nature, or as sick, that is, biologically unwell), but has also been used as a major argument of progressive gay movements, where the fundamental argument has been – if it's natural, how can it be wrong? Magnus Hirschfeld, the leading spokesperson of the German gay movement before 1933, argued, with clinical data and photos to back it up, that gay people were physically different from heterosexual masculine men and feminine women (see Steakley, 1975). This view was equally implicit in the early declaration of the Chicago Society for Human Rights, almost certainly the first gay rights organisation in the USA, founded in 1924, which refers to 'people . . . of mental and physical abnormalities' (Katz, 1976, p.385) in defining gay people. The more successful Mattachine Society declared in its 1950 prospectus:

> *We, the androgynes of the world, have formed this responsible corporate body to demonstrate by our efforts that our physiological and psychological handicaps need be no deterrent in integrating 10% of the world's population towards the constructive social progress of mankind* (ibid., p.410).

Even when gay people have rejected such notions of themselves, there is still a sense of how ambiguously we are placed in relation to gender and often a welcoming of that as a release from the rigidity of the sex roles. Michael Bronski (1978, p.205) in his article on gay men's fondness for stars such as Garland, Davis, Streisand and so on, links this implicitly to a notion of the refinement of gay sexuality:

> *The most refined form of sexual attractiveness (as well as the most refined form of sexual pleasure) consists of going against the grain of one's sex . . . It is not surprising that in a society which places so much emphasis upon gender roles gay men should be drawn to personalities that blur such distinctions.*

The idea then of gender androgyny expressing homosexuality, or being appealing to homosexual people, is deeply ingrained in straight and gay consciousness alike. Thus although Garland did not express 'sexual androgyny' (in the sense of homosexuality),* it was enough that she so regularly expressed 'gender androgyny'.

*The evidence for Garland's having had a lesbian relationship is slight and only mentioned by Finch (1975, p.129–30).

One of Garland's most popular records (of a song never used in her films) is *In-between.* In this she bemoans, comically, her status as an 'in-between'. Recorded soon after *Dear Mr Gable (You Made Me Love You),* what the teenage singer is in between is childhood and adulthood and this is certainly the burden of the lyrics. Yet the song, written by Roger Edens especially for her, does also hint that to be a teenage girl is to be a gender misfit, to have no appropriate role – 'Too old for toys/Too young for boys' – and it draws on the idea of unattractiveness discussed above and which can clearly be seen to relate to ideas of achieving femininity:

My dad says I should bother more
About my lack of grammar
The only thing that bothers me
Is my lack of glamour.

In any event, the song, especially given the fact that it is sung by Judy Garland, has a potential gay resonance strong enough to be used by David Clough for his BBC television play **Belles** (transmitted 27.5.83) about a drag act performing, through an error of booking, at a straight-laced night-club. The play explored the way that drag might be seen as a challenge to straight society, but it reaches that point via the doubts the main character Michael (Martyn Hesford) has about himself. At one moment in this earlier part he listens to Garland singing *In-between;* he has on drag make-up but no wig, thus underlining the gender confusion of his act; and he listens in close-up to the verse which, taken (as here) without the chorus that follows, can easily be assumed to be about homosexuality or drag:

Fifteen thousand times a day
I hear a voice within me say
Hide yourself behind a screen
You shouldn't be heard
You shouldn't be seen
You're just an awful in-between

That's what I am
An in-between

It's just like small pox quarantine
I can't do this
I can't go there
I'm just a circle in a square
I don't fit in anywhere.

Images of Garland as androgynous go right back to the earliest part of her career. A still from **The Big Revue** (1929), when she was still Frances Gumm, shows her with her sisters, unmistakably little girls but wearing playful versions of male clothing – top hats (with polka dots reminiscent of minstrel garb), tuxedo fronts and cuffs. Similarly, she wore a sailor dress for promotion pictures with Jackie Cooper in 1936, overalls in contrast to Deanna Durbin in **Every Sunday** (1936), dungarees in **Everybody Sing** (1938) and a cowgirl outfit in **Girl Crazy** (1943), and so on. Yet in context none of these is remarkable. A certain androgyny has always been permissible for women in fashion, in chorus girl costumes, in the tomboy role. The markers of masculinity in such cases – a hat, a collar, a cuff – are never so strong as the markers of femininity – bosom, skirt, exposed stockinged legs. It is an interesting fact that the assumption of a little masculinity is a standard feature of woman-as-spectacle.

The Gumm sisters. Judy Garland (l.), then Frances Gumm, in **The Big Revue** (1929): 'playful versions of male clothing' (photo Lester Glassner)

Frequently the allusion is not to adult masculinity but to boyishness – the sailor suit refers to a standard item of boys' wear in the early part of the century, the tomboy connotes the notion of a pre-adolescent phase. Although one could no doubt reread these standard androgynous images in a gay way, it seems unnecessary – such images are probably too normalised really to stand out.

The one exception is perhaps her role in Fanny Brice's Baby Snooks routine in the final show in **Everybody Sing.** Garland comes on dressed in a short-trousered velvet suit. 'Is you a girl or boy?' asks Brice, dressed in a baby's nightie. 'They call me Little Lord Fauntleroy,' replied Garland, evading the question. 'What's a Fauntleroy?' counters Brice, and the rest of the number plays around with this, with Judy never able to say whether she is male or female. Although a comic routine, Garland performs it with an increasingly tremulous and frustrated quality as it develops.

Judy Garland and Fanny Brice in **Everybody Sing** (1938)

Judy Garland in cowgirl outfit in **Girl Crazy**

Judy Garland sings *Get Happy* in **Summer Stock**

Everbody Sing is very much a vehicle film, tailored to the images of all its stars. Although only her second full role in a feature, the Garland character is from the start put down as lovable but unattractive ('mother's little ugly duckling'), and with her first in dungarees for *Down on Melody Farm* and then done up as Little Lord Fauntleroy the film rather insistently allies this unattractiveness to in-betweenism. It is, however, only later in the career that the androgyny of Garland goes much further, and in two opposing directions – stylish and tragi-comic, vampish and trampish.

Get Happy from **Summer Stock** represents the former, and yields one of the most often reproduced images of Garland. It is not only the outfit itself that creates the stylish androgyny, but Garland's movement. Near the beginning she pushes the hat from behind with her palm over her brow, a gesture taken from Apache dancing, suggesting the male going in for the sexual kill (cf. Maurice Chevalier in **Love Me Tonight).** Towards the end she pushes it back from the front with her thumb, in a gesture more

Maurice Chevalier in **Love Me Tonight** (1932)

reminiscent of James Cagney getting down to business. Garland's relationship to the male dancers is ambivalent too. She is centred by them and this, plus her stockinged legs, insists on her femininity, but they do not surround and present her as other male choruses do in musical numbers centred on a female star. They are choreographed in a balanced (but not uniform) style around her, and her dancing picks up on the movements of different men at different times. In other words, she is to some degree 'one of the boys', especially in a movement of flexing the thighs forward and heels up that is used for men in urban ballets of

Judy Garland in **A Star is Born** (1954)

the kind Jerome Robbins developed. As Jane Feuer has pointed out (1982, p.119), this image from her last number in an MGM film is picked up by her first number in her next film (but four years later and post-1950) **A Star is Born.** Costuming and routine are similar, and the sense of being one of the boys is heightened by her amused and professional handling of Norman who stumbles drunkenly on stage during the number, *You Gotta Have Me Go With You.* Yet the number is preceded by the shot that introduces her in the film; she is seen from behind adjusting her stockings, a standard glamour introductory shot. In *Get Happy* and *You Gotta . . .* she is always both glamorous, sexy and one of the boys, a kind of androgyny not dissimilar to that of Dietrich and Garbo. It is an androgynous image with sex appeal.

This sort of image was used in the concerts and on television, but even more characteristic were the tramp and clown outfits derived from *A Couple of Swells* in **Easter Parade** and *Be A Clown* in **The Pirate.** There is an echo of it in the ragamuffin look in *Lose That Long Face* in **A Star is Born,** a look used in **Inside Daisy Clover** in evident reference to Garland. Her man's shirt and tights for *Somewhere There's a Someone* in **A Star is Born** is less evidently within this style, but the narrative context (cheering up Norman by evoking the absurdity of the movie world from which he is now excluded) suggests a tragi-comic element, and the outfit

Fred Astaire and Judy Garland as *A Couple of Swells* in **Easter Parade**

has yielded a particularly memorable off-the-set portrait in this particular androgynous mode.

It was the concerts and television shows that most used the tramp look. The first Palace show in 1951 rechoreographed *A Couple of Swells* and placed it late in the show so that Garland was still in its costume for the final *Over the Rainbow*. This pattern was reworked in all the subsequent stage shows. The concerts, which were not in this sense staged, also often ended with Garland seated on the edge of the stage in a spotlight that lit only her face which, with its cropped hair style, could evoke the tramp image. Whereas the vamp-androgyne is an image that emphasises sexuality, the tramp-androgyne dissolves both sexuality and gender. Its use equally for comic and sentimental numbers puts it straight in the line of the classic tragi-comic image of the clown, an evocation of the show business heritage very appropriate for Garland's 'Miss Show Business' tag. If in the vamp gay men could identify with someone whose sexuality is accepted by the boys, in the tramp we could identify with someone who has left questions of sexuality behind in an androgyny that is not so much in-between (marked as both feminine and masculine) as without gender. The loose clothing (though it is male) conceals the shape of the body; and the abstraction of the image of just a face in a spotlight completes the dream of escape from sex role.

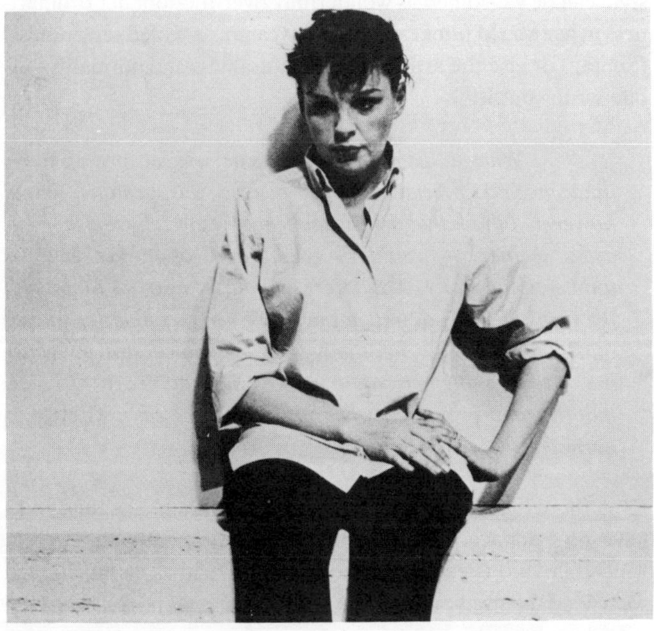

Judy Garland (photo by Bob Willoughby)

Camp

Thirdly, Judy Garland is camp. Several people have tried to define both what camp is and its relationship to the situation and experience of gay men. (I do not propose to re-enter the fray of whether it is politically and culturally progressive – for, see Cohen and Dyer, 1980; against, see Britton, 1978/79.) It clearly is a defining feature of the male gay subculture. Jack Babuscio (1977) suggests that it is the fact of being able to pass for straight that has given gays the characteristically camp awareness of surfaces, of the social constructedness of sex roles (see also Russo, 1979 and Dyer, 1977). Mark Booth (1983) stresses gay men's sense of marginality which is turned into an excessive commitment to the marginal (the superficial, the trivial) in culture. Either way, camp is a characteristically gay way of handling the values, images and products of the dominant culture through irony, exaggeration, trivialisation, theatricalisation and an ambivalent making fun of and out of the serious and respectable. (See, in addition to those already cited, Sontag, 1964 and Boone, 1979, especially the latter's discussion of 'trivialisation'.)

The object of camp's making fun is often a star like Bette Davis or Shirley Bassey, and Garland can be read like that. She is imitable, her appearance and gestures copiable in drag acts (e.g. Jim Bailey, Craig Russell in **Outrageous!**); her later histrionic style can be welcomed as wonderfully over-the-top; her ordinariness in her MGM films can be seen as camp, as 'failed seriousness' (Sontag) or else the artificiality of naturalness and normality – as one writer put it:

> *What I liked about this star was a sort of naive innocence, a sweetness that I believe was genuine. Now, however, I think that the gay audience – especially the sophisticated segment – would or might look upon her style as unintentional camp. This effect, of course, also has to do with the kinds of roles, as well as the movie vehicles, that she played in. Because her style, her stance, was so damned 'straight' (I use the word meaning 'serious' or 'sober'), it amounted to, as I said, that of camp or at least what appeared to be camp* (Letter to author).

Even this writer doesn't say that he reads Garland like this, and I have no evidence to suggest that it is the predominant way in which Garland is camp. Anybody can be read as camp (though some lend themselves to it more readily than others), but Garland is far more inward with camp. She is not a star turned into camp, but a star who expresses camp attitudes.

In the later years this was clear, particularly in the chat at the concerts and (specially composed) verses of the songs. At Carnegie Hall, she leads into *San Francisco* with the verse:

I never will forget
Jeanette MacDonald
Just to think of her
It gives my heart a pang
I never will forget
How that brave Jeanette
Just stood there
In the ruins
And sang
A-a-a – and sang –

MacDonald was well established as camp queen of an already pretty camp genre, operetta. Garland extends the beat on a hum before singing her name and pauses for three waves of laughter before proceeding – for that audience, it was enough to mention MacDonald's name to get a camp response. The send-up (of MacDonald singing in the debris of an earthquake in the film **San Francisco**) that follows serves for those who haven't already got the point that MacDonald is camp.

At her last appearance at the Talk of the Town (like the Carnegie Hall concert, available on record), she introduces *I'd Like to Hate Myself in the Morning:*

> *I have to do a new song. I haven't been taught a new song since Clive Brooks was a girl. (Laughter) Was he?*

Here the gay connection is more direct – Brooks is not only camp because of his high, clipped upper-class English accent, but because his gender can be called into question. Her 'Was he?' presumably is a question about Brooks' sexuality, to which gay cognoscenti in the audience perhaps knew the answer.

Garland's reputation for being camp (rather than being seen as camp) was reinforced by stories that were published after her death. Her own awareness of the gay connection is made clear. Barry Conley in **Gay News** quotes (as have others) Liza Minnelli quoting her mother – '"When I die I have visions of fags singing *Over the Rainbow* and the flag at Fire Island being flown at half mast"' (Conley, 1972, p.11), Fire Island being a largely gay beach resort near New York. Christopher Finch (1975, pp.129–

30) writes at length of her ease in the predominantly gay milieu of the Freed unit while Brad Steiger (1969, pp.103–4) talks of her frequenting gay bars in New York in the sixties. One person who wrote to me told me that her

> *last professional engagement in New York City was when she took money under the table singing in a lesbian bar on East 72nd Street called 'Sisters' in about 1968–9. This $50 per session or so was paid to support her whenever she would 'wander in' and pretend to sing 'impromptu' so she could support her drug habit. Very sad but nonetheless true* (Letter to author).

Although published after her death, these accounts do suggest how versed she was, or could be assumed to be, in gay culture, how inward with its procedures and cadences.

The effect of her camp is to act back on her films, in two ways. First, a repertoire of stories has built up around her own attitude to her work. The more these become part of gay fan talk, the more they will inform perception of the films. The simple sparkle of *Who (Stole My Heart Away)?* from **Till the Clouds Roll By** (1946) becomes ironic when one knows Garland's apparent mirth at singing the song when she was several months pregnant. Her role as the sweet, out-of-town girl in **Easter Parade** may look more acerbic when informed by what Charles Walters says she said to him before shooting:

> *Look sweetie, I'm no June Allyson, you know. Don't get cute with me. None of that batting-the-eyelids bit, or the fluffing the hair routine for me, buddy!* (quoted in Finch, 1975, p.158).

The knife edge between camp and hurt, a key register of gay culture, is caught when one takes together her intense performance cf the scene after Norman's death in **A Star is Born** and her remark quoted by George Cukor, who had expressed amazement that she had reproduced such intensity over two long takes:

> *Oh, that's nothing. Come over to my house any afternoon. I do it every afternoon. But I only do it once at home* (ibid., p.197).

Similarly one may hear the pathos of the later performances of *Over the Rainbow* differently when one has at the back of one's mind Garland's remark to a fan who begged her never to 'forget the rainbow' – 'Why, madam . . . how could I ever forget the rainbow? I've got rainbows up my ass' (Conley, 1972, p.10 – the source of the story is Liza Minnelli).

But, secondly, this bringing together of Garland's reported remarks and the films is not necessary, because there is camp in the text of the films themselves. Her films with Vincente Minnelli, especially *The Great Lady Gives an Interview* in **Ziegfeld Follies** (1946) and **The Pirate** (1948), are rather obviously camp pieces, with their elements of theatricality, parody and obvious artifice. Garland's camp humour often has the effect of sending up the tone and conventions of her films. In **Presenting Lily Mars** (1943) she sings *When I Hear Beautiful Music* in a night-club, a scene in which the character's star quality is demonstrated. It is camp in two ways. First, Garland mocks the operetta style singing of Marta Eggerth, who has sung the song earlier in the film. This is a standard camp Garland routine, used on radio broadcasts in the late thirties and forties as well as in other films – she uses excessively elaborated trills, oversweetened notes and hand-wringing, shoulder-rolling, lip-curling gestures to summon up the pretty-pretty soprano style of Jeanette MacDonald, Deanna Durbin and other such stars. Secondly, Garland's performance also sends up this standard girl-gets-big-break-in-night-club. As a waiter flashes by with a tray held aloft, Garland's eyes roll, she backs away from him, then smirks at this detail of restaurant *mise-en-scène* that would go by unnoticed in a more straight performance. There are several other such moments where Garland draws attention to the way the number has been set up, culminating in her stumbling over the inconveniently placed drum set, undercutting any notion of this as a straightforward moment of star quality triumphant.

In such ways Garland can seem to be reflecting back either on her own image in the film or on the vehicle in which she has been placed. As Wade Jennings (1979, p.324) puts it, there is 'lurking in her eyes and the corners of her mouth . . . [a] suppressed mirth . . . [that] threatens to mock the silly plot and the two-dimensional character she plays'. When Dorothy emerges from the house into Oz for the first time, she says to her dog, 'Toto, I don't think we're in Kansas anymore.' It's obviously a funny line, given the shimmering Oz sets, but is it said in a knowing, camp way? Are we laughing directly at Dorothy's charming naivety, or with Garland at the over-the-top sets and Dorothy's artful gingham frock? Impossible to determine, of course, though Christopher Finch (1975, p.85) does point out that, at 17, Garland in **The Wizard of Oz** is 'an adolescent with a grown-up's singing voice *acting* the part of a child' – the possibility of ambivalence, play, fun with the part and plot is at least there. In

Toto, I don't think we're in Kansas any more

1982, Rockshots, a gay greetings card company, issued a card depicting Garland as Dorothy, in gingham with Toto in a basket, in a gay bar, with her opening line in Oz as the message inside. It is not just the incongruity of juxtaposing Dorothy/Garland and men in gay macho style clothes and poses that is camp; the card has picked on Garland's irony towards Dorothy and Oz, an irony very easily transposed by gay men to their cultivation of an exaggeratedly masculine style and scene. Just as Garland in Oz can be seen as both in the magic world and yet standing outside it too, so gay culture is ambivalent about its construction of a fantasy scene that is both keenly desired and obviously a put-on.

For the most part Garland's campness might be seen as mildly sabotaging her roles and films. (Those who dislike camp feel that it is in fact deeply destructive in its insistently making fun of everything – see Britton, 1978/79.) It is seldom of a piece with the rest of the film, and only **The Pirate** seems to use Garland's campness in a sustained fashion in its play with sex roles and spectacular illusion, two of the standard pleasures musicals offer.

Sex role send-up centres on the male in **The Pirate** and Gene Kelly's role and performance as Serafin. Garland as Manuela functions in two ways in relation to this. First, she is given a string of lines to deflate his machismo, whether it be his corny chatting up of her at their first meeting or his pretending to be the pirate Macoco later in the film. In the first case she directly mocks his lines; in the second, when she has just discovered that he is not really Macoco, she wildly exaggerates the excitement of his virility while at the same time humorously hitting where it really hurts him, Serafin, by saying what a lousy actor he is (and thus calling forth a display of bruised male pride, in itself a standard moment in Gene Kelly films). Since Manuela is also a sweet, vulnerable girl, yearning for her dreams to come true (in other words, the standard Garland role), these salty quips unsettle the easy acceptance of that girlish image; and as a lot of the humour is underscored by the way Garland delivers the lines, it can seem like the performer wittily intervening to deconstruct the characters she is playing and playing opposite.

Secondly, in terms of narrative, Garland/Manuela is placed as the subject of desire; that is, the film is about her desire for a truly exciting life, and man. She constructs this desire in the image of the pirate Macoco, Mack the Black, that she has learnt from her story books. The film plays on this desire, at the same time playing with the Garland image. It turns out that, as in **The Wizard of Oz, Meet Me in St Louis** and so many of her films she is indeed engaged to Macoco, only neither she nor anyone else in the town knows it. The problem is that the real Macoco is nothing like the story book Mack the Black – he is the fat, fussy mayor of the town. Garland/Manuela has her heart's desire right at hand – only he doesn't look like her heart's desire at all, a point brought out later by the governor-general's rather lascivious comparison of Serafin (whom he thinks is Macoco) and the usual dull round of real pirates. When, in their betrothal scene, Macoco (alias Don Pedro) tells Garland/Manuela that he has no plans to take her travelling, that home is best, there is a reaction shot of her uttering, appalled, the word 'home', the exact reversal in tone of Dorothy/Garland's line at the end of **The Wizard of Oz** or Esther/Garland's at the end of **Meet Me in St Louis.**

The person who does look like her heart's desire, but who she knows can only play the part of it, is the actor Serafin. Through the film's complex narrative peripeteia Manuela/Garland is led to the point where she knowingly settles for the illusion of her heart's desire – she settles for Serafin.

In addition to this narrative progression, the film also suggests the degree to which Manuela's fantasies are themselves socially constructed fictions. Under hypnosis (which close-ups clearly indicate we are meant to take as having really occurred), Manuela expresses her true feelings to Serafin – but these turn out to be an amalgam of lines from earlier in the film, taken from either her story book on Macoco or from what Kelly/Serafin has said to her when he tries to chat her up. So from the book we get 'Someday he'll swoop down on me like a chicken hawk and carry me away', from Serafin's spiel 'Beneath this prim exterior there are depths of emotion, romantic longings'. Garland hardens her voice for such lines, to give them a dryly comic edge; yet her performance of going into hypnosis is soft and tremulous, her performance of the ensuing *Mack the Black* deliriously all-out (with wildly swinging camera movements to underscore this). This kind of moving in and out of the 'emotional truth' of the character and situation allows the film, allows Garland, to point to both the vivid intensity of repressed feelings and the fact that those feelings are themselves culturally constructed, not a given authenticity.

In the second hypnosis scene, Garland/Manuela pretends to be hypnotised in order to flush out Don Pedro (whom she has only just realised is Macoco) and save Serafin (who is about to be hanged as Macoco). Again she plays out the exaggerated longing for his virility as she did before when mocking Serafin's impersonation of Macoco; here it is actually on Serafin's stage, thus doubly underlining this declaration of desire as performance. With the way it is played and cut-ins of Don Pedro getting more and more frustrated that this love is not being directed at him, the real Macoco, the whole scene works several levels of camp together. However, when Garland/Manuela then sings *Love of My Life* to Kelly/Serafin/?Macoco, although still on stage the film goes into a soft-focus close-up and standard heartfelt crooning from Garland – in other words, at the point in the film most signalled as illusion, we get the most direct expression of 'true' feeling. It is in the recognition of illusion that camp finds reality.

The treatment of men as spectacle reiterates this. It is Kelly not Garland who is the centre of the big production numbers, *Niña* and *The Pirate Ballet,* each of them emphasising sex through costuming (tights and shorts, respectively), Kelly's movement (for instance, wiggling his bottom at the groups of women clustered about him in *Niña,* flexing his thighs in *The Pirate Ballet)* and camerawork (sinuous camera movement in the first, low angle, crotch centred positioning in the second). *Niña* is not observed by Garland/Manuela, but generally in the film man/Kelly as spectacle is established as being from her point of view. (In itself quite unusual in Hollywood films – **Rebecca,** for instance, a film whose first half hour is entirely constructed around the never named female protagonist's desire for Maxim/ Laurence Olivier, nonetheless entirely denies this character any point of view shots of him.) The film keeps shifting its/our perspective on how we are to take Garland/Manuela's libidinous looking. The first introduction of the fantasy of Mack the Black is through the story book pictures of him. The film starts with this, with us directly enjoying the coloured drawings of him (in which he is predominantly constructed as a figure of rape fantasy). It is only after a few pages of drawings have been turned that the camera draws back and we discover we have been entering the fantasy through Manuela's point-of-view. *The Pirate Ballet* starts from a dissolve of her looking at Serafin in the street from her bedroom window, the implication being that the wild, exploding, sexy number that follows is how she sees him in her mind's eye. Later, when she is making fun of his assumption of Macoco's machismo, she runs her eyes up and down him saying with relish, 'Let me look my full at you'. In the second hypnosis scene, she circles Kelly/Serafin (pretending she thinks he's Macoco), gazing at him as she celebrates (sends up) his masculinity, while Don Pedro/Macoco is cut-in looking on, desiring to be looked at by Manuela as she is now looking at Serafin. The film thus fully allows Kelly as sex object, to a more sustained degree than any male star between Rudolph Valentino and John Travolta; and at the same time plays around with him as spectacle, so that he is both turn-on and send-up.

The film's resolution is the acceptance of both together, the embrace of the illusion of spectacle, and in the widest sense. Not only does Garland/Manuela settle for someone who only looks like (is the spectacle of) Mack the Black, she also opts to become a player – she will get to travel the world through the

Judy Garland and Gene Kelly in *Be a Clown* from **The Pirate**

profession of pretence. Again the film gives another twist to the paradox. In the final number, *Be a Clown,* Garland/Manuela and Kelly/Serafin perform as clowns; right at the end in close-up they look at each other and burst into laughter. There is no reason why Manuela and Serafin, in love, doing a jolly number, should not as characters burst into laughter; but many have seen this as Judy Garland and Gene Kelly falling about laughing at the fun they've just been making in the film. Thus dressed as clowns on a stage in a very 'theatrical' movie-musical (in terms of performance and *mise-en-scène*), the 'real' attitude of the performers is held to come through – the reality of the pretence of illusion.

In lines and role, as well as the way Garland plays them, **The Pirate** explores a camp attitude towards life. Play on illusion and reality does not have to be seen as camp or gay. As throughout this chapter, I am neither claiming that only gay men could see it this way or that these aspects need be understood as camp or gay. What I am saying is that the particular way in which **The Pirate** plays with questions of artifice, together with the presence of Garland in it and her particular way of delivering a line as funny, make it particularly readable within the gay male subcultural discourse of camp.

I Could Go On Singing

Judy Garland's last film, **I Could Go On Singing,** made in England in 1962, is her most gay film. It is clearly aware of the gay audience and is in many ways a summation of the gay way of reading her image.

It is a gay film in relatively explicit ways. To begin with, there is a gag that would have little reverberation to anyone who had never heard of the New York gay community's favourite seaside spot – when drunk, Jenny/Judy remarks, 'I've had enough to float Fire Island.' (This was probably lost on most of the contemporary British gay audience too.) Secondly, the casting of Dirk Bogarde at this point in his career must have been suggestive. **Victim** had been released two years before and although he presented his decision to appear in this film (a thriller campaigning for reform in the laws relating to homosexuality) as social consciousness rather than coming out as gay, his image was indelibly coloured with homosexuality thereafter. In the light of it, earlier films such as **The Spanish Gardener** and **The Singer Not the Song** look much gayer, centring as they do on an intense relationship between, respectively, a man and a boy, and a bandit and a priest. Since both **Victim** and **I Could Go On Singing** he has

Judy Garland and Montgomery Clift on the set of **Judgement at Nuremberg** (1961)

played a number of gay or crypto-gay roles, including **The Servant, Modesty Blaise** and **Death in Venice.** In addition, his appearance puts him in line with an important gay stereotype of the period, that of the 'sad young man'. There is a similarity between his looks and those of another gay identification star, Montgomery Clift – Garland and Clift were both in **Judgement at Nuremberg** and although not on screen together, an off-screen photograph and reports on the filming emphasised the affinity between these two 'unhappy' stars. Bogarde and Clift are both like other visual and literary representations of the 'sad young man' (see Dyer, 1983), so that they were available to be read through this particular version of how gay men were seen and saw themselves.

Bogarde as star has all this gay resonance even apart from the way **I Could Go On Singing** uses him. In no way does the film suggest his character, David, is homosexual, but it is worth considering the fact that the film centres on a non-sexual relationship between a man and a woman both played by stars with a particular significance for gay culture. The film has come up with a plot situation akin to the well-documented syndrome of intimate relationships between gay men and straight women. It even seems to toy with the exploration of some of the ambiguities of this. Jenny/Garland and David/Bogarde have had a sexual relationship in the past, producing a son, Matt, whom Jenny agreed to disown. In the film Jenny wants to reclaim Matt and this is the drive of the narrative, but it does seem as if part of that could involve renewing her sexual relationship with David. Towards the end, when David is coaxing her out of her drunkenness to go back to the Palladium for her show, he says to her, 'I've always loved you', and she says, nodding, 'That's where it ends?' This is the clearest point at which the woman's desire for greater physical intimacy is blocked by the man's resolute refusal. It is a moment that gay men might flatter themselves could easily happen to them, and it might be doubly flattering in that not only does it express female desire for them but also shows their absolute power to refuse it. This reading has to be hedged about with more than the usual number of qualifiers; but the casting of Garland *and* Bogarde in this very unusual (for a film story) kind of relationship does make such a reading at the very least possible.

Stars and basic plot situation (plus one gag) give one a lead into a gay reading of the film, and this is facilitated by the degree to which it accords with most of those aspects of the Garland image discussed above – emotionality, gay sensibility, androgyny, camp. I'll discuss these in a moment. The one aspect that is less explored is ordinariness. As mentioned above, the basic situation – international showbiz celebrity tries to (re)integrate self into ordinariness of family life – starts off from this motif in Garland's image. But the man is Bogarde, equally readable as gay and therefore not the stuff of normal family life (at any rate, as that is represented; it is statistically common); and, as we'll see, the treatment of the public school sequence renders it anything but nice and normal.

As a vehicle film for Judy Garland, **I Could Go On Singing** both provides several big emotional musical numbers (*By Myself, Hello Bluebird, It Never Was You, I Could Go On Singing),* at least two set pieces of untramelled (one-take) emotional expressivity (listening to her son telling her he does not want to go to Paris with her; talking out her problems drunkenly to David), and a plot that subjects her to filial and uxorial rejection and yet has her come smiling through by the end. The songs in particular work a kind of meteorological tradition for Garland, going back to *Over the Rainbow* and all setting up something to do with the weather that you have to experience in order to reach happiness – for instance, *I'm Always Chasing Rainbows* in **Ziegfeld Girl,** *In the Valley Where the Evening Sun Goes Down* in **The Harvey Girls,** *Look for the Silver Lining* in **Till the Clouds Roll By,** *Friendly Star* in **Summer Stock,** *It's a New World* in **A Star is Born.** In **I Could Go On Singing,** *Hello Bluebird* recalls the line from *Over the Rainbow,* 'If happy little bluebirds fly'; *It Never Was You* opens with 'I've been running through rain and the winds that follow after'; and the title songs ends with 'I must keep on singing like a lark going strong with my heart on the wings of a song singing day'.

Songs, set pieces and plot place centrally (as the pre-1950 films did not) the suffering and courage nexus so important to the gay reading of Garland. *By Myself* (which had been featured on the album **Alone** as well as in concerts) catches it most fully, not only because it is a song about courage in the face of loneliness and rejection, but because it occurs immediately after the scene in which Matt has overheard Jenny and David rowing about him and realised that Jenny is the mother who had deserted him. Jenny/Garland sings *By Myself* as a fierce refusal to be brought low by Matt's horror that this nice and famous lady friend of his father is in fact his despised mother. The way it is performed also builds up from a more fragile, vulnerable opening to a stronger, tougher climax. At the start she is picked out by a spotlight that makes her look small and isolated in the centre of the stage (and of the scope screen); her hands hang by her side, a frail gesture from someone associated with a more histrionic repertoire of gestures; there is a thin, slightly sinister string accompaniment and she comes in on the song without any intro, thus her voice is unsupported, musically exposed. For much of the first time through (all sung slowly) she is in close-up, which enables us to see a remarkable facial gesture on the words

I'll face the unknown
I'll build a world of my own

where she both backs away with her head (as if from adversity) and yet hardens her facial features (as if setting them against 'the unknown'). The second time through is up tempo, with full orchestra and belting delivery. She struts about the stage or stands with feet apart, one hand on her hip. Where the first time through 'I'll build a world of my own' has the wistfulness of Dorothy's yearning for Oz, the second time it has the determination and confidence of *Get Happy* in **Summer Stock** or *Rockabye Your Baby with a Dixie Melody* from the concerts. At the end she does not wait for the applause, but walks straight off flicking back the flimsy tab curtain as she does. Thus we are to take it that the emotion of the song is carried over into the singer herself, so overcome with emotion that she cannot relate to the audience at all or else, more literally, turning her back on her audience as well as her family to be utterly alone. (In the subsequent narrative she gets drunk and is contemptuous of the audience David says is waiting for her.)

This overlap of performance and life not only authenticates the former (she – Jenny/Garland – 'really feels' the emotion she sings because it is an emotion from her life), but also fits into the film's treatment of the theatricality of experience, part of Jack Babuscio's definition of the gay sensibility. He suggests that in **A Star is Born** and especially **I Could Go On Singing**

> *Garland took on roles so disconcertingly close to her real-life situation and personality that the autobiographical connections actually appeared to take their toll on her physical appearance from one scene to the next* (Babuscio, 1977, p.46).

Be that as it may, the film not only shows the 'real life' of the narrative being performed out in the stage numbers, but also in part theatricalises the narrative too. Jenny/Garland's first appearance is delayed – we see her from behind getting out of a taxi and going up to a door; we then cut to inside the house, a maid answering the door and Jenny/Garland entering and stepping into the light. In part this is just the kind of heavy-handed approach to movies (and hence to big stars) characteristic of British cinema, but it does have the effect of making Jenny/Garland's first appearance in the narrative into a theatrical event. Later the scenes in Jenny's Palladium dressing room suggest a continuity between theatre and life. There is, for instance, a tracking shot on to a close-up of a microphone in a spotlight which immediately cuts on Jenny/Garland dramatically pulling back a curtain and appearing in her stage clothes. The effect of the cut is to make it seem that Jenny/Garland is making her stage entrance, but in fact the second shot is in the dressing room. Life there is lived in a theatrical manner, and this use of the curtained-off area in the room occurs elsewhere in the film. I do not wish to give the impression that **I Could Go On Singing** has the allusive play on theatre and reality that **The Pirate** or **A Star is Born** have, merely to indicate that something of that interplay is present in it, reinforced by the use of androgyny and camp.

Garland is not herself presented in an androgynous manner in the film. She wears dresses, or skirts with bolero jackets and blouses, throughout, and her hair is done in a conventional sixties perm. Her strutting or feet astride performance style in some of the numbers might be seen as mannish, but there is certainly no attempt to build on this (as compared to outfits and performance style in *You Gotta Have Me Go With You* and the finale of *Born in a Trunk* in **A Star is Born**). What is present is a bizarre sequence at Matt's public school. The idea of Judy Garland going to a rugby match in high heels, as she does here, is already outlandish enough, but it is far outstripped by her attendance at the school production of Gilbert and Sullivan's **HMS Pinafore.** As it is an all-male school, the female parts are played by boys and the sequence we see is in fact Matt dressed as a girl singing happily about the virtues of the British sailor (say no more). After the show, we are introduced to the dressing room with a shot of Matt from behind taking off his skirt and bustle before turning round – the camera position is a standard voyeuristic one (cf. **The Prince and the Showgirl** dressing room sequence discussed in the Monroe chapter), but the undressing reveals the theatrical construction of gender and the figure's turning round reminds us that what looked like a girl is really a boy. Jenny/Garland comes in and joins in with the 'boys', in varying stages of undress, and the 'girls', to varying degrees in drag and made-up as women. They sing *His Sisters and His Cousins and His Aunts* together, with Jenny/Garland interpolating skat snatches, joking with the boys and in every way suggesting she is completely at home with this androgynous crew. The sequence ends with Matt trying to persuade David to let him go to see her show. Matt tries to act winningly to get David's permission; he is in boy's clothes but still wearing green eyeshadow, with his full mouth emphasised by lipstick; when he turns to Jenny and says, 'Make him say yes', he smiles coyly back at David. Matt is their son and that is how we're supposed to think of him;

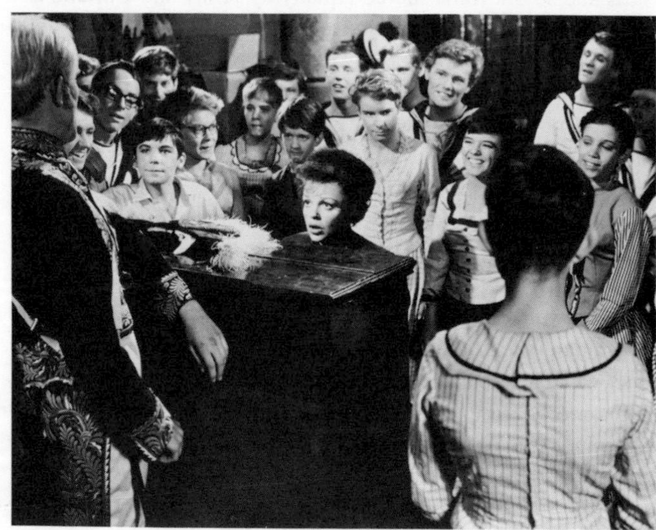

Judy Garland at the piano in **I Could Go On Singing**

but there is something insistently gay about Judy Garland and Dirk Bogarde tussling over so emphatically androgynous a creature. When later Judy and Dirk row about Matt, the scene as written and played feels like they are squabbling over the same man – only the context makes it clear that it is a boy, their son, not a love object.

One can get a lot of camp fun out of all this weird gender play, and Garland is also herself camp in her performance. One might read this deconstructively, her performance foregrounding the facticity of the vehicle, but oddly her camp seems to reinforce the basic realist/illusionist style of the film. What Jenny/Garland is camp about is herself and her situation, but not the artifices of the film itself. For instance, there is an awkward moment when Jenny first meets Matt and she remarks that he looks like David, his father; Matt believes he is adopted and so thinks this cannot be possible, but David adds that nonetheless he could resemble him – 'adopted children do sometimes grow like their parents, like dogs'. Jenny/Garland nods encouragingly along with this, but double takes and echoes archly, 'Like dogs?' This could be taken to undercut the after all rather laborious script here, but it seems easier to take it as sending up David's po-faced awkwardness and the incongruity of her own situation as an incognito parent.

Such small interjections, reaction shots, gestural and vocal inflections run throughout Garland's performance in the film. It's not a very funny script, but she plays it as if it is a stream of wit. Characteristically for camp humour, much of this is directed against herself – her renowned unreliability, drunkenness and so on, traits shared equally by Jenny and Garland and mentioned within the first five minutes of the film. This can also be putting herself down as a woman. At one point in their first conversation, David asks her if she still knits. 'Oh, no,' she replies, 'nothing I knitted ever fitted.' The joke is partly about the stiltedness of this conversation with a man she still loves who is the father of the son she wants to see; but it also bespeaks her lack of feminine skills. She delivers it seated in a wing-backed upright chair, with her left hand raised and resting on one of its wings. It is an elegant pose that is coded as appropriate to a certain kind of graceful woman – but Jenny/Garland does not look like that kind of woman, she is adopting a mannered elegance much as a man does when he is being camply effeminate. Line and pose together are camp because she is not a graceful woman with traditional feminine skills. We may find it funny, but like much camp humour there is a sour edge when you start to think about it. Garland, like gay men, makes fun of herself as she fails to be womanly. Unlike gay men, however, Garland does not have the patriarchal position to retreat to that, in **I Could Go On Singing,** Dirk Bogarde does. This is the point at which Garland's image outruns the gay reading of her, where the fact of her being a woman, not a gay man, takes her image in a whole lot of other directions.

One of the letters I received while researching this chapter tells an interesting story.

> *When I was 13 or 14 the whole 3rd form at school voted what film they would like to see at the end of term. My friend and I liked Garland and we wanted to see* **The Wizard of Oz.** *We were labelled as 'poofs' and also laughed at for being childish, unlike many other 3rd formers who thought they were so mature because they wanted to see a sex film.* **Dr No** *was the film they finally chose.*
>
> *I have found out since that my friend was gay a couple of years later and I find it interesting . . . that 2 gays, unaware of each other's sexual preferences remained in solidarity for Garland. Furthermore, while we were not conscious of her as a gay person's film star, it is interesting to note that we were brutally labelled as 'poofs' for our choice of film* (Letter to author).

This letter suggests that the gay reading of Judy Garland was not just something that gay men would pick up as they entered the gay scene; it suggests rather that a person identifying themselves as gay (or probably 'different') would intuitively take to Garland as an identification figure.

There need be no consciousness of the connection. Another letter I received stressed that the writer had had a 'teenage love affair with MGM's great star' and at the same time

> *I was a gay teenager and well aware of it. In fact I was very active. I did not in any way connect the gay life with the affinity to Judy* (Letter to author).

Only retrospectively does he see 'the reasons for gays being attracted to Judy', which he understands in terms of her appeal to 'sensitive males'.

What seems to be happening in such cases is a coming together of two homologous structures – a star image with strong elements of difference within ordinariness, androgyny and camp, and a way of interpreting homosexual identity that is widely available in society in both dominant and subcultural discourses. The gay subculture would develop the most elaborated, the most inward of readings of Garland, would pick up on the nuances and inflections of her image that could be read in a gay way, and it is this that I have been exploring. But the classmates of the letter writer above clearly sensed, without probably having any knowledge of the composition of Garland's audience, that there was something about Garland which chimed with their sense of what 'poofs' were, a connection between image and social identity that the writer himself also made intuitively.

There is nothing arbitrary about the gay reading of Garland; it is a product of the way homosexuality is socially constructed, without and within the gay subculture itself. It does not tell us what gay men are inevitably and naturally drawn to from some in-built disposition granted by their sexuality, but it does tell us of the way that a social–sexual identity has been understood and felt in a certain period of time. Looking at, listening to Garland may get us inside how gay men have lived their experience and situation, have *made* sense of them. We feel that sense in the intangible and the ineffable – the warmth of the voice, the wryness of the humour, the edgy vigour of the stance – but they mean a lot because they are made expressive of what it has been to be gay in the past half century.

References

Preface

BFI Education (1982) **Star Signs** (London: British Film Institute).

Dyer, Richard (1979a) **Stars** (London: British Film Institute).

Dyer, Richard (1979b) **The Stars, Teachers Study Guide no. 1** (London: British Film Institute).

Introduction

Allyson, June with Spatz-Leighton, Frances (1982) **June Allyson** (New York: G.P. Putnam).

Arnold, Eve (1976) **The Unretouched Woman** (New York: Knopf).

1: Monroe and Sexuality

Alanen, Antti (1982) **Marilyn Monroe** (Painatus: Valtion painatuskeskus).

Alpert, Hollis (1956) 'Sexual Behaviour in the American Movie', **Saturday Review,** no. 39, 23 June 1956.

Brinkmann, Paul Delbert (1971) 'Dr Alfred C. Kinsey and the Press', unpublished PhD. dissertation, Department of Mass Communications, University of Indiana.

Brownmiller, Susan (1975) **Against Our Will** (London: Secker & Warburg).

Conway, Michael and Ricci, Mark (1964) **The Films of Marilyn Monroe** (Secaucus NJ: Citadel).

Cook, Pam (1979/80) 'Star Signs', **Screen,** vol. 20, no.3/4, Winter 1979/80, pp.80–8.

Dennis, Norman, Henriques, Fernando and Slaughter, Clifford (1969) **Coal Is Our Life** (London: Tavistock; originally published Eyre & Spottiswoode, 1956).

Dyer, Richard (1979) **The Dumb Blonde Stereotype** (London: British Film Institute).

Ellman, Mary (1970) **Thinking About Women** (New York: Harcourt, Brace Jovanovich).

Farnham, Marynia and Lundberg, Ferdinand (1947) **Modern Woman, the Lost Sex** (New York: Harper & Bros).

Foucault, Michel (1980) **The History of Sexuality** (New York: Vintage Books).

French, Brandon (1978) **On the Verge of Revolt** (New York: Frederick Ungar).

French, Marilyn (1978) **The Women's Room** (London: André Deutsch).

Friedan, Betty (1963) **The Feminine Mystique** (New York: W.W. Norton).

Fuller, Peter (1980) **Art and Pyschoanalysis** (London: Writers & Readers).

Gagnon, J.H. and Simon, William (1974) **Sexual Conduct** (London: Hutchinson).

Gornick, Vivian and Moran, Barbara K. (eds) (1971) **Woman in Sexist Society** (New York: Basic Books).

Guiles, Fred Lawrence (1969) **Norma Jean** (New York: McGraw-Hill).

Harris, Thomas B. (1957) 'The Building of Popular Images: Grace Kelly and Marilyn Monroe', **Studies in Public Communication,** 1.

Haskell, Molly (1974) **From Reverence to Rape** (New York: Holt, Rinehart & Winston).

Hess, Thomas B. (1972) 'Pinup and Icon', in Hess, Thomas B. and Nochlin, Linda (eds) **Woman as Sex Object** (New York: Newsweek).

Mailer, Norman (1973) **Marilyn** (London: Hodder & Stoughton).

Maskosky, Donald R. (1966) 'The Portrayal of Women in Wide Circulation Magazine Short Stories 1905–1955', unpublished PhD dissertation, University of Pennsylvania.

McIntosh, Mary (1976) 'Sexuality', **Papers on Patriarchy** (Lewes: Women's Publishing Collective).

Metallious, Grace (1957) **Peyton Place** (New York: Frederick Muller).

Metallious, Grace (1960) **Return to Peyton Place** (New York: Frederick Muller).

Miller, Douglas T. and Nowak, Mary (1977) **The Fifties, the Way We Really Were** (Garden City, New York: Doubleday).

Morantz, Regina Cornwell (1977), 'The Scientist as Sex Crusader, Alfred C. Kinsey and American Culture', **American Quarterly,** vol. XXIX, no.5.

Ryan, Mary P. (1975) **Womanhood in America** (New York: Franklin Watts).

Saxton, Martha (1975) **Jayne Mansfield and the American Fifties** (Boston: Houghton Mifflin).

Sheinwold, Patricia Fox (1980) **Too Young to Die** (London: Cathay).

Sinclair, Marianne (1979) **Those Who Died Young** (London: Plexus).

Spence, Jo (1978/79) 'What do people do all day?' **Screen Education,** Winter 1978/79, no.29, pp.29–45.

Stanley, Liz (1977) **The Problematic Nature of Sexual Meanings** (Manchester, British Sociological Association Sexuality Study Group).

Steinem, Gloria (1973) 'Marilyn – The Woman Who Died Too Soon', **The First Ms. Reader** (New York: Warner).

Tickner, Lisa (1978) 'The Body Politic: Female Sexuality and Women Artists Since 1970', **Art History,** vol. 1, no.2, June 1978, pp.236–241.

Trilling, Diana (1963) 'The Death of Marilyn Monroe', **Claremont Essays,** (New York: Harcourt, Brace Jovanovich).

Walters, Margaret (1978) **The Nude Male** (New York and London: Paddington Press).

Weeks, Jeffrey (1981) **Sex, Politics and Society** (London and New York: Longman).

Whitman, Howard (1962) **The Sex Age** (Indianapolis: Bobbs Merrill).

Zolotow, Maurice (1961) **Marilyn Monroe, An Uncensored Biography** (London: W.H. Allen).

2: Paul Robeson: Crossing Over

Baldwin, James (1978) **The Devil Finds Work** (London: Pan; Michael Joseph 1976).

Berghahn, Marion (1977) **Images of Africa in Black American Literature** (Totawa NJ: Rowman & Littlefield).

Bogle, Donald (1974) **Toms, Coons, Mulattoes, Mammies and Bucks** (New York: Bantam; Viking, 1973).

Bordman, Gerald (1980) **Jerome Kern, His Life and Music** (Oxford: Oxford University Press).

Clark, Kenneth (1960) **The Nude** (London: Penguin)

Clum, John M. (1969) 'Ridgely Torrence's Negro Plays: A Noble Beginning', **South Atlantic Quarterly,** LXVIII, pp.96–108.

Cripps, Thomas (1970) 'Paul Robeson and Black

Identity in American Movies', **Massachusetts Review,** vol. 11, no.3, Summer 1970, pp.468–85.

Cripps, Thomas (1977) **Slow Fade to Black** (Oxford: Oxford University Press).

Cruse, Harold (1969) **The Crisis of the Negro Intellectual** (London: W.H. Allen).

Cruse, Harold (1978) 'The Creative and Performing Arts and the Struggle for Identity and Credibility', in Harry A. Johnson (ed.) **Negotiating the Mainstream** (Chicago: American Library Association).

Cuney-Hare, Maude (1936) **Negro Musicians and their Music** (Washington DC: Associated Publishers).

Dodge, Mabel (1936) **Intimate Memoirs,** Volume II (New York: Harcourt Brace).

Dyer, Richard (1982a) **'The Son of the Sheik', The Movie,** no. 126, pp.2512–13.

Dyer, Richard (1982b) 'Don't Look Now', **Screen,** vol.23, nos.3/4 pp. 61–73.

Ellis, John (1982) 'Star/Industry/Image', in BFI Education, **Star Signs** (London: British Film Institute) pp.1–12.

Embree, Edwin R. (1945) **Thirteen Against the Odds** (New York: Viking).

Frederickson, George M. (1972) **The Black Image in the White Mind** (New York: Harper & Row).

Freedomways Editors (eds) (1978) **Paul Robeson, the Great Forerunner,** (New York: Dodd, Head).

Friedberg, Anne (1980–1) 'Approaching *Borderline*', **Millennium,** nos. 7/8/9, Fall/Winter 1980–1, pp.130–9.

Goldstein, Malcolm (1974) **The Political Stage** (New York: Oxford University Press).

Haggard, Rider (1958) **King Solomon's Mines** (Harmondsworth: Penguin; originally published 1885).

Hamilton, Virginia (1974) **Paul Robeson: The Life and Times of a Free Black man** (New York: Dell).

Harrington, Ollie (1978) 'Our Beloved Pauli', in *Freedomways* Editors, qv, pp.100–6.

Henderson, Edwin Bancroft (1939) **The Negro in Sports** (Washington DC: Associated Publishers).

Huggins, Nathan Irvin (1971) **Harlem Renaissance** (New York: Oxford University Press).

Isaacs, Edith (1947) **The Negro in the American Theater** (New York: Theater Arts).

Johnson, James Weldon (1968) **Black Manhattan** (New York: Atheneum; originally published 1930).

Johnson, James Weldon and Johnson, J. Rosamond (1969) **The Books of American Negro Spirituals** (New York: Viking; originally published 1925 and 1926).

Kempton, Murray (1955) **Part of Our Time** (New York: Simon & Schuster).

Klein, Michael (1975) '*Native Land:* Praised Then Forgotten', **Velvet Light Trap,** no.14, Winter 1975, pp.12–16.

Levine, Lawrence L. (1977) **Black Culture and Black Consciousness** (Oxford: Oxford University Press).

Locke, Alain (ed.) (1968) **The New Negro** (New York: Johnson Reprint; originally published 1925).

Mannin, Ethel (1930) **Confessions and Impressions** (New York: Doubleday, Doran) pp.157–61 ('Paul Robeson: Portrait of a Great Artist').

Miers, Earl Schenck (1942) **Big Ben** (Philadelphia: Westminster Press).

Nizhny, Vladimir (1962) **Lessons with Eisenstein,** trans. Ivor Montague and Jay Leyda (London: Allen & Unwin).

Noble, Peter (1948) **The Negro in Films** (London: Skelton Robinson).

Ovington, Mary White (1927) **Portraits in Color** (New York: Viking) pp.205–15.

Pines, Jim (1975) **Blacks in Films** (London: Studio Vista).

Polhemus, Ted and Proctor, Lynn (1978) **Fashion and Anti-Fashion** (London: Thames & Hudson).

Raleigh, John Henry (1965) **The Plays of Eugene O'Neill** (Carbondale and Edwardsville: Southern Illinois University Press).

Roach, Hildred (1973) **Black American Music Past and Present** (New York: Crescendo).

Roberts, John Storm (1973) **Black Music of Two Worlds** (London: Allen Lane).

Robeson, Eslanda Goode (1930) **Paul Robeson, Negro** (New York: Harper & Bros).

Robeson, Paul (1978) 'The Culture of the Negro', in *Freedomways* Editors q.v. pp.65–7 (originally published in **The Spectator,** 15 June 1934, vol. 152, pp.916–17).

Rosenberg, Marvin (1961) **The Masks of Othello** (Berkeley and Los Angeles: University of California Press).

Schlosser, Anatol I. (1970) 'Paul Robeson, His Career in the Theatre, in Motion Pictures, and on the Concert Stage', unpublished PhD. dissertation, New York University.

Sergeant, Elizabeth Shepley (1926) 'The Man with his

Home in a Rock: Paul Robeson', **New Republic** 3 March, 1926, pp.40–4.

Seton, Marie (1958) **Paul Robeson** (London: Dennis Dobson).

Southern, Eileen (1971a), **The Music of Black Americans** (New York: W.W. Norton).

Southern, Eileen (ed.) (1971b) **Readings in Black American Music** (New York: W.W. Norton).

Stowe, Harriet Beecher (1981) **Uncle Tom's Cabin** (Harmondsworth: Penguin; originally published 1852).

Taylor, Karen Malpede (1972) **People's Theatre in America** (New York: Drama Book Specialists).

Walton, Ortiz (1972) **Music Black White and Blue** (New York: William Morrow).

Williams, Raymond (1973) **The Country and the City** (London: Chatto & Windus).

3: Judy Garland and Gay Men

Anger, Kenneth (1981) **Hollywood Babylon** (New York: Dell; originally published Paris, J.J. Pauvert, 1959).

Babuscio, Jack (1977) 'Camp and the Gay Sensibility', in Richard Dyer (ed.) **Gays and Film** (London: British Film Institute) pp.40–57.

Bathrick, Serafina (1976) 'The Past As Future: Family and the American Home in *Meet Me in St. Louis*', **The Minnesota Review,** New Series 6, Spring 1976, pp. 7–25.

Boone, Bruce (1979) 'Gay Language as Political Praxis, the Poetry of Frank O'Hara', **Social Text,** no. 1, pp.59–92.

Booth, Mark (1983) **Camp** (London: Quartet).

Brinson, Peter (1954) 'The Great Come-Back', **Films and Filming,** December 1954, p.4.

Britton, Andrew (1977) '*Meet Me in St Louis:* Smith or The Ambiguities', **Australian Journal of Screen Theory,** no. 3, pp.7–25.

Britton, Andrew (1978/79) 'For Interpretation, Against Camp', **Gay Left,** no.7, Winter 1978/79, pp.11–14.

Bronski, Michael (1978) 'Judy Garland and Others, Notes on Idolization and Derision', in Karla Jay and Allen Young (eds) **Lavender Culture,** (New York: Harcourt Brace Jovanovich).

Cohen, Derek and Dyer, Richard (1980) 'The Politics of Gay Culture', in Gay Left Collective (eds) **Homosexuality, Power and Politics** (London: Allison & Busby) pp.172–86.

Conley, Barry (1972) 'The Garland Legend: The Stars Have Lost Their Glitter', **Gay News,** no. 13, pp.10–11.

Di Orio, Al (1975) *Little Girl Lost: The Life and Hard Times of Judy Garland* (New York: Manor Books).

Dyer, Richard (1977) 'Its Being So Camp As Keeps Us Going', **Body Politic,** no. 36, September 1977.

Dyer, Richard (1977) 'Entertainment and Utopia', **Movie,** no.24, pp.2–13.

Dyer, Richard (1977) 'Four Films of Lana Turner', **Movie,** no.25, pp.30–52.

Dyer, Richard (1982) '*A Star is Born* and the Construction of Authenticity', in BFI Education, **Star Signs** (London: British Film Institute) pp.13–22.

Dyer, Richard (1983) 'Seen to be Believed, Some Problems in the Representation of Gay People as Typical', **Studies in Visual Communication,** vol. 9, no.2, Spring 1983, pp.2–19.

Feuer, Jane (1982) **The Hollywood Musical** (London: Macmillan/British Film Institute).

Finch, Christopher (1975) **Rainbow, the Stormy Life of Judy Garland** (London: Michael Joseph).

Goldman, William (1969) **The Season** (New York: Harcourt Brace & World Inc.).

Gramann, Karola and Schlüpmann, Heide (1981) 'Unnatürliche Akte. Die Inszenierung des Lesbischen im Film', in Karola Gramann *et al.,* **Lust und Elend: das erotische Kino** (Munich & Luzern: Bucher) pp.70–93.

Greig, Noel and Griffiths, Drew (1981) **As Time Goes By** (London: Gay Men's Press).

Howard, Dumont (1981) 'The Garland Legend', **Blueboy,** January 1981.

Jennings, Wade (1979) 'Nova: Garland in *A Star is Born*', **Quarterly Review of Film Studies,** Summer 1979, pp.321–37.

Katz, Jonathon (ed.) (1976) **Gay American History** (New York: Thomas Y. Cromwell).

Marcuse, Herbert (1964) **One Dimensional Man** (Boston: Beacon Press).

Meyers, Janet (1976) 'Dyke Goes to the Movies', **Dyke,** Spring 1976.

Ruehl, Sonja (1982) 'Inverts and Experts: Radclyffe Hall and the Lesbian Identity', in Rosalind Brunt and Caroline Rowan (eds), **Feminism, Culture and Politics** (London: Lawrence & Wishart).

Russo, Vito (1979) 'Camp', in Martin P. Levene (ed.), **Gay Men, the Sociology of Male Homosexuality** (New York: Harper & Row pp.205–10; originally published **The Advocate,** 19 May, 1976).

Russo, Vito (1980/81) 'Poor Judy', **Gay News,** no.205, 11 December–7 January 1980–81 (supplement), pp.14–15.

Sheldon, Caroline (1977) 'Lesbians and Film: Some Thoughts', in Richard Dyer (ed.) **Gays and Film** (London: British Film Institute), pp.5–26.

Sontag, Susan (1964) 'Notes on Camp', **Partisan Review,** XXXI, no.4.

Steakley, James (1975) **The Homosexual Emancipation Movement in Germany** (New York: Arno).

Stearn, Jess (1962) **The Sixth Man** (New York: Macfadden Books).

Steiger, Brad (1969) **Judy Garland** (New York: Ace Books).

Watney, Simon (1980) 'The Ideology of GLF', in Gay Left Collective (eds), **Homosexuality, Power and Politics** (London: Allison & Busby), pp.64–76.

Weeks, Jeffrey (1977) **Coming Out: Homosexual Politics from the Nineteenth Century to the Present** (London: Quartet).

Whitaker, Judy (1981) 'Hollywood Transformed', **Jump Cut** nos. 24/25, pp.33–35.

Wood, Robin (1976) **Personal Views** (London: Gordon Fraser).

Woodcock, Roger (1969) 'A Star is Dead', **Jeremy,** vol.1, no.1, 1969, pp.15–17.

Index